Probability and Mathematical Statistics (Continued)

MUIRHEAD • Aspects of Multivariate Statistical Theory
PARZEN • Modern Probability Theory and Its Applications
PURI and SEN • Nonparametric Methods in General Linear Models
PURI and SEN • Nonparametric Methods in Multivariate Analysis
RANDLES and WOLFE • Introduction to the Theory of Nonparametric Statistics
RAO • Linear Statistical Inference and Its Applications, *Second Edition*
RAO and SEDRANSK • W.G. Cochran's Impact on Statistics
ROHATGI • An Introduction to Probability Theory and Mathematical Statistics
ROHATGI • Statistical Inference
ROSS • Stochastic Processes
RUBINSTEIN • Simulation and The Monte Carlo Method
SCHEFFE • The Analysis of Variance
SEBER • Linear Regression Analysis
SEBER • Multivariate Observations
SEN • Sequential Nonparametrics: Invariance Principles and Statistical Inference
SERFLING • Approximation Theorems of Mathematical Statistics
TJUR • Probability Based on Radon Measures
WILLIAMS • Diffusions, Markov Processes, and Martingales, Volume I: Foundations
ZACKS • Theory of Statistical Inference

Applied Probability and Statistics

ABRAHAM and LEDOLTER • Statistical Methods for Forecasting
AGRESTI • Analysis of Ordinal Categorical Data
AICKIN • Linear Statistical Analysis of Discrete Data
ANDERSON, AUQUIER, HAUCK, OAKES, VANDAELE, and WEISBERG • Statistical Methods for Comparative Studies
ARTHANARI and DODGE • Mathematical Programming in Statistics
BAILEY • The Elements of Stochastic Processes with Applications to the Natural Sciences
BAILEY • Mathematics, Statistics and Systems for Health
BARNETT • Interpreting Multivariate Data
BARNETT and LEWIS • Outliers in Statistical Data, *Second Edition*
BARTHOLOMEW • Stochastic Models for Social Processes, *Third Edition*
BARTHOLOMEW and FORBES • Statistical Techniques for Manpower Planning
BECK and ARNOLD • Parameter Estimation in Engineering and Science
BELSLEY, KUH, and WELSCH • Regression Diagnostics: Identifying Influential Data and Sources of Collinearity
BHAT • Elements of Applied Stochastic Processes, *Second Edition*
BLOOMFIELD • Fourier Analysis of Time Series: An Introduction
BOX • R. A. Fisher, The Life of a Scientist
BOX and DRAPER • Evolutionary Operation: A Statistical Method for Process Improvement
BOX, HUNTER, and HUNTER • Statistics for Experimenters: An Introduction to Design, Data Analysis, and Model Building
BROWN and HOLLANDER • Statistics: A Biomedical Introduction
BUNKE and BUNKE • Statistical Inference in Linear Models, Volume I
CHAMBERS • Computational Methods for Data Analysis
CHATTERJEE and PRICE • Regression Analysis by Example
CHOW • Econometric Analysis by Control Methods
CLARKE and DISNEY • Probability and Random Processes: A First Course with Applications, *Second Edition*
COCHRAN • Sampling Techniques, *Third Edition*
COCHRAN and COX • Experimental Designs, *Second Edition*
CONOVER • Practical Nonparametric Statistics, *Second Edition*

CONOVER and IMAN • Introduction to Modern Business Statistics
CORNELL • Experiments with Mixtures: Designs, Models and The Analysis of Mixture Data
COX • Planning of Experiments
DANIEL • Biostatistics: A Foundation for Analysis in the Health Sciences, *Third Edition*
DANIEL • Applications of Statistics to Industrial Experimentation
DANIEL and WOOD • Fitting Equations to Data: Computer Analysis of Multifactor Data, *Second Edition*
DAVID • Order Statistics, *Second Edition*
DAVISON • Multidimensional Scaling
DEMING • Sample Design in Business Research
DILLON and GOLDSTEIN • Multivariate Analysis: Methods and Applications
DODGE • Analysis of Experiments with Missing Data
DODGE and ROMIG • Sampling Inspection Tables, *Second Edition*
DOWDY and WEARDEN • Statistics for Research
DRAPER and SMITH • Applied Regression Analysis, *Second Edition*
DUNN • Basic Statistics: A Primer for the Biomedical Sciences, *Second Edition*
DUNN and CLARK • Applied Statistics: Analysis of Variance and Regression
ELANDT-JOHNSON and JOHNSON • Survival Models and Data Analysis
FLEISS • Statistical Methods for Rates and Proportions, *Second Edition*
FLEISS • The Design and Analysis of Clinical Experiments
FOX • Linear Statistical Models and Related Methods
FRANKEN, KÖNIG, ARNDT, and SCHMIDT • Queues and Point Processes
GALAMBOS • The Asymptotic Theory of Extreme Order Statistics
GIBBONS, OLKIN, and SOBEL • Selecting and Ordering Populations: A New Statistical Methodology
GNANADESIKAN • Methods for Statistical Data Analysis of Multivariate Observations
GOLDSTEIN and DILLON • Discrete Discriminant Analysis
GREENBERG and WEBSTER • Advanced Econometrics: A Bridge to the Literature
GROSS and CLARK • Survival Distributions: Reliability Applications in the Biomedical Sciences
GROSS and HARRIS • Fundamentals of Queueing Theory, *Second Edition*
GUPTA and PANCHAPAKESAN • Multiple Decision Procedures: Theory and Methodology of Selecting and Ranking Populations
GUTTMAN, WILKS, and HUNTER • Introductory Engineering Statistics, *Third Edition*
HAHN and SHAPIRO • Statistical Models in Engineering
HALD • Statistical Tables and Formulas
HALD • Statistical Theory with Engineering Applications
HAND • Discrimination and Classification
HILDEBRAND, LAING, and ROSENTHAL • Prediction Analysis of Cross Classifications
HOAGLIN, MOSTELLER and TUKEY • Exploring Data Tables, Trends and Shapes
HOAGLIN, MOSTELLER, and TUKEY • Understanding Robust and Exploratory Data Analysis
HOEL • Elementary Statistics, *Fourth Edition*
HOEL and JESSEN • Basic Statistics for Business and Economics, *Third Edition*
HOGG and KLUGMAN • Loss Distributions

(*continued on back*)

Beyond ANOVA, Basics of Applied Statistics

Beyond ANOVA, Basics of Applied Statistics

RUPERT G. MILLER, JR.
STANFORD UNIVERSITY

JOHN WILEY & SONS
New York • Chichester • Brisbane • Toronto • Singapore

Library of Congress Cataloging in Publication Data:

Miller, Rupert G.
Beyond ANOVA, basics of applied statistics.

Bibliography: p.
Includes indexes.
1. Mathematical statistics. I. Title. II. Title:
Beyond A.N.O.V.A., basics of applied statistics.

QA276.M473 1985 519.5 84-19511
ISBN 0-471-81922-0

Printed in the United States of America

10 9 8 7 6 5 4 3 2

PREFACE

These are the confessions of a practicing statistician. They expose to public view what I am likely to do with a set of data. I may therefore live to regret setting pencil to paper. Yet there does not seem to be a book that tells a student how to attack a set of data. There are books on the analysis of variance, there are books on nonparametric statistics, there are books on this and that, but which technique should I use on the data? This book attempts to go beyond any specific discipline and consider the variety of techniques that can be brought to bear on a problem. The statistical problem is the central focus, not a particular theoretical approach.

This book is written for M.S. and Ph.D. students of statistics who have some knowledge of the analysis of variance, nonparametric statistics, etc., but who are still unclear on what to do when confronted with data. It is hoped that this book will be useful as well to biologists, social scientists, and engineers who know some statistics and want to handle their own data analysis.

It will be immediately apparent that this book in no way covers the complete range of statistical problems and ideas. Designs more complex than the two-way classification (e.g., three-way classifications and Latin squares) are not included, nor is multiple regression. The hope is that the reader will grasp the basic ideas behind the simpler analyses and thus understand how to cope with the more complex situations. Unmentioned are problems where the basic random variables are binary valued or categorical. Also, no attempt has been made to incorporate the techniques of multivariate analysis or time series analysis.

Since the statistical techniques based on normal theory have been so central to the development and teaching of statistics, the structure of each chapter (or subchapter) is to first present the normal theory methods and then investigate what happens when the normality assumption and other assumptions break down. In most chapters, this leads to sections on nonnormality, unequal variances, and dependence.

Exercises are included at the end of each chapter. Some are theoretical, and others involve data analysis. The latter were selected for their relevance and interest from my files of projects at the Stanford Medical Center.

Rupert G. Miller, Jr.

London, England

Stanford, California

ACKNOWLEDGMENTS

The writing of this book was begun during 1972-73 while I was on sabbatical leave in London, England. Peter Armitage at the London School of Hygiene and Tropical Medicine and David Cox at the Imperial College of Science and Technology graciously supplied me with office space and access to library facilities. A fellowship from the John Simon Guggenheim Memorial Foundation provided financial support for the academic year. Gordon Ray and the staff of the Guggenheim Foundation were very supportive and generously increased the stipends for Fellows living abroad when the U.S. dollar was devalued in the middle of the year. From the Statistics and Probability Program at the National Science Foundation under the leadership of Bob Agins, I received support during the summer of 1973 for continued work on the book.

Since 1974 my research has been supported by a grant from the National Institute of General Medical Sciences. The Program Administrators for this grant have been Margaret Carlson and Americo Rivera. Undoubtedly, some time spent on this book can be ascribed to the grant.

By 1976 three chapters of the book had been written, but they were put on the shelf to gather dust at the start of 1977 when I became Editor of the *Annals of Statistics*. The chapters were removed from the shelf for a one quarter sabbatical leave in the spring of 1981 again in London at Imperial College. Unfortunately, by this time the already written chapters were badly in need of revision to bring them up to date.

A number of skilled technical typists have worked on this manuscript over the years. Judi Davis, Carolyn Knutsen, Gail Lammond, and Martha Thomas all typed parts of the original first three chapters. The revised and later chapters were typed by Judi Davis, Karola Decleve, and Nora Kundo. To Karola Decleve fell the task of making numerous changes and corrections to produce a final manuscript.

Judi Davis did a superb job of typesetting the finished manuscript with the TeX typesetting system on our departmental VAX computer. Wiley then photographed the pages to produce this book.

Valuable comments were received from Bill Brown, Naihua Duan, Barbara Miller, and Lincoln Moses, who read the original chapters. The students in Statistics 233A,B, Applied Statistics, at Stanford University were subjected to the revised chapters as a text during Autumn and Winter Quarters 1983–84. They found a number of misprints and suggested various improvements. Particularly notable in this regard were Jim Cutler, John Willett, and Gary Williamson.

Barbara Miller provided encouragement and assistance throughout. Especially helpful was her rescue of the references when the Apple computer file in which I was accumulating the references became so large that the word processor started scrambling some of them. Also, she was of great assistance in helping to proofread and index the entire book.

To all the aforementioned individuals I express my appreciation and thanks. I hope that this book is worthy of their support and efforts.

R. G. M.

CONTENTS

1 **ONE SAMPLE** **1**

1.1 Normal Theory 1
1.2 Nonnormality 5
1.2.1 Effect 5
1.2.2 Detection 10
1.2.3 Correction 16
Transformations 16
Nonparametric Techniques 19
Robust Estimation 28
1.3 Dependence 32
1.3.1 Effect 34
1.3.2 Detection 35
1.3.3 Correction 36
Exercises 37

2 **TWO SAMPLES** **40**

2.1 Normal Theory 40
2.2 Nonnormality 41
2.2.1 Effect 41
2.2.2 Detection 44
2.2.3 Correction 44
Transformations 44
Nonparametric Techniques 45
Robust Estimation 54
2.3 Unequal Variances 56
2.3.1 Effect 56
2.3.2 Detection 58
2.3.3 Correction 58
Transformations 58
Other Tests 60
2.4 Dependence 63

Exercises 64
3 ONE-WAY CLASSIFICATION 67
FIXED EFFECTS 69
3.1 Normal Theory 69
3.1.1 Analysis of Variance (ANOVA) 69
3.1.2 Multiple Comparisons 71
3.1.3 Monotone Alternatives 76
3.2 Nonnormality 80
3.2.1 Effect 80
3.2.2 Detection 82
3.2.3 Correction 82
Transformations 82
Nonparametric Techniques 82
Robust Estimation 89
3.3 Unequal Variances 89
3.3.1 Effect 89
3.3.2 Detection 92
3.3.3 Correction 92
3.4 Dependence 94
RANDOM EFFECTS 95
3.5 Normal Theory 95
3.5.1 Estimation of Variance Components . . 96
3.5.2 Tests for Variance Components . . . 99
3.5.3 Estimation of Individual Effects . . . 101
3.5.4 Estimation of the Overall Mean . . . 104
3.6 Nonnormality 105
3.6.1 Effect 105
3.6.2 Detection 107
3.6.3 Correction 108
3.7 Unequal Variances 109
3.8 Dependence 110
Exercises 111

4 **TWO-WAY CLASSIFICATION** **117**
FIXED EFFECTS 118
4.1 Normal Theory 118
4.1.1 Analysis of Variance (ANOVA) . . . 119
4.1.2 Multiple Comparisons 129
4.1.3 Monotone Alternatives 131
4.2 Nonnormality 135
4.2.1 Effect 135
4.2.2 Detection 136
4.2.3 Correction 137
Transformations 137
Nonparametric Techniques 137
Robust Estimation 140
4.3 Unequal Variances 140
4.3.1 Effect 140
4.3.2 Detection 141
4.3.3 Correction 141
4.4 Dependence 141
MIXED EFFECTS 143
4.5 Normal Theory 143
4.6 Departures from Assumptions 149
RANDOM EFFECTS 150
4.7 Normal Theory 150
4.7.1 Estimation of Variance Components . . . 151
4.7.2 Tests for Variance Components 155
4.7.3 Estimation of Individual Effects and Overall Mean 157
4.8 Departures from Assumptions 158
Exercises 159

5 **REGRESSION** **164**
REGRESSION MODEL 168
5.1 Normal Linear Model 168
5.1.1 One Sample: General Intercept 168

5.1.2 One Sample: Zero Intercept 181
5.1.3 Multisamples: General Intercepts 184
5.1.4 Multisamples: Zero Intercepts 191
5.2 Nonlinearity 193
5.2.1 Effect 194
5.2.2 Detection 194
5.2.3 Correction 196
5.3 Nonnormality 199
5.3.1 Effect 199
5.3.2 Detection 202
5.3.3 Correction 202
5.4 Unequal Variances 207
5.4.1 Effect 208
5.4.2 Detection 209
5.4.3 Correction 210
5.5 Dependence 214
ERRORS-IN-VARIABLES MODEL 220
5.6 Normal Theory 224
5.7 Departures from Assumptions 231
Exercises 234

6 RATIOS 241

6.1 Normal Theory 242
6.2 Departures from Assumptions 249
Exercises 254

7 VARIANCES 259

7.1 Normal Theory 259
7.2 Nonnormality 264
7.2.1 Effect 264
7.2.2 Detection 265
7.2.3 Correction 266
7.3 Dependence 275
Exercises 276

REFERENCES 279
AUTHOR INDEX 308
SUBJECT INDEX 313

Beyond ANOVA, Basics of Applied Statistics

Chapter 1

ONE SAMPLE

The simplest problem is that of a sample from a single population where the aim of the statistical analysis is to estimate, or test a hypothesis about, the location of the population. Many of the techniques for detecting and correcting departures from assumptions are illustrated in this basic setting.

1.1. Normal Theory.

Let $y_1, ..., y_n$ be independently distributed as $N(\mu, \sigma^2)$.* For hypothesis testing the null hypothesis is $H_0 : \mu = \mu_0$ and the alternative could be either one-sided $H_1 : \mu > \mu_0$ or two-sided $H_1 : \mu \neq \mu_0$. From the estimation point of view the problem is to estimate μ and construct a confidence interval for it.

The variables y_i may themselves be combinations of other variables. For instance, when observations u and v are taken on subjects paired to eliminate the effect of nuisance variables, y_i may be the paired difference $u_i - v_i$ for the ith pair, and the null hypothesis of no difference has $\mu_0 = 0$. Or, in a different setting, the ratio $y_i = u_i/v_i$ may be the natural variable in which case $\mu_0 = 1$ might be the null hypothesis.

The likelihood ratio test of $H_0 : \mu = \mu_0$ vs. $H_1 : \mu \neq \mu_0$ leads

---

* "$N(\mu, \sigma^2)$" denotes a normal distribution with mean μ and variance σ^2.

to *Student's* (1908) *t statistic*

$$t = \frac{\bar{y} - \mu_0}{s/\sqrt{n}}, \tag{1.1}$$

which has a Student's t distribution with $n - 1$ df.*,** Theoretical hypothesis testing says reject the null hypothesis if $|t| > t_{n-1}^{\alpha/2}$, where $t_{n-1}^{\alpha/2}$ is the upper 100 $(\alpha/2)$ percentile of the t distribution and α is the preselected significance level.

A formal hypothesis testing framework is conceptually very useful and has led to great advances in statistical theory. However, I don't remember ever having fixed α and having tested a hypothesis. Instead, I report the *P value*, which is the probability under the null hypothesis of obtaining a result equal to, or more extreme than, the observed. In this case $P = 2P\{t_{n-1} > |t|\}$, where t_{n-1} has a t distribution on $n - 1$ df and t is the observed value of the statistic (1.1). P is a measure of the credibility of the null hypothesis. The smaller P is, the less likely one feels the null hypothesis can be true. For discussion of the P value see Gibbons and Pratt (1975) and Pratt and Gibbons (1981, Chapter I, Section 4).

Bayesian statisticians would report a different measure of the credibility of the null hypothesis, namely, the posterior probability of its being correct. However, this requires knowing the prior probability of the null hypothesis being true and the probability measure over the alternative hypotheses. I am never fortunate enough to know these. DeGroot (1973) has tried to bring the P value and Bayesian philosophy closer together by giving examples in which the P value can be interpreted as a posterior probability.

Believers in likelihood would report the entire likelihood function. I have been involved in situations where calculating the likeli-

---

* $\bar{y} = \sum_{i=1}^{n} y_i/n$; $s^2 = \sum_{i=1}^{n}(y_i - \bar{y})^2/(n-1)$.

** "df" denotes degrees of freedom.

hood function was informative and helpful. It indicates which alternatives are compatible with the data and which are not. However, it involves more work than computing a P value – a function must be tabled or a graph drawn. Also, it requires the assumption of a parametric model. For routine scientific reporting, the P value is simpler and is more nearly universally understood by scientific investigators.

Some might argue that even with the P value classical hypothesis testing is being practiced because statements such as "$P \leq .05$" or "$P \leq .01$" will appear in scientific articles and results are not published unless $P \leq .05$. I would say that the inclusion of statements like "$P \leq .05$" is more a result of imprecision or extensive tables being unavailable rather than hypothesis testing with $\alpha = .05$ being practiced. Also, I rarely make a more refined statement like "$P = .001$" because, except for certain nonparametric distributions, the accuracy of such a statement depends on an assumption about the form of the distribution very far out in the tails. Robustness of far out tails of a distribution is not easily guaranteed and reports such as "$P = .001$" may be overly optimistic.

It cannot be denied that many journal editors and investigators use $P \leq .05$ as a yardstick for the publishability of a result. This is unfortunate because not only P but also the sample size and the magnitude of a physically important difference determine the quality of an experimental finding. For an experiment the sample size may be necessarily small due to limitations of time and/or money, and a finding with $P = .10$ may be far more striking than a result in another paper which has $P = .05$ but much larger sample size. The larger the sample size the smaller P has to be to warrant attention. This involves the power of the test and the probability of detecting small differences of no practical worth. Differences can be highly statistically significant and yet be of such small magnitude as to have no practical significance. Also, with large sample sizes the analysis may be detecting a small bias in the experiment rather than a true difference.

The t statistic (1.1) can be used as a pivotal statistic to construct a confidence interval on μ_0:

$$\mu_0 \in \bar{y} \pm t_{n-1}^{\alpha/2}\, s/\sqrt{n}. \tag{1.2}$$

Often in the scientific literature a do-it-yourself confidence interval is reported. Namely, the mean $\bar{y}$ and the *standard error* $s/\sqrt{n}$ are presented, sometimes with a "±" sign between them. Armed with these, the sample size, and a t table one could construct (1.2), but most times a rough mental calculation of the mean plus and minus two standard errors suffices for the reader. Similarly, in graphs the custom is to plot the mean as a point and a vertical line whose extent measures plus and minus one standard error (see Figure 1.1). Unfortunately, I have the feeling that most readers unconsciously construe the vertical line to be the 90%, 95%, or 99% confidence interval on μ_0.

Occasionally, when the intent is to convey the variability of the data, the vertical line will denote plus and minus one standard deviation. More sophisticated plots, called *box-and-whisker plots*, can be used to describe the variability in the data. For details see Tukey (1977).

The t statistic (1.1) can also be used to test the one-sided alternative $H_1 : \mu > \mu_0$. In this case the P value is $P = P\{t_{n-1} > t\}$. The corresponding one-sided confidence interval is

$$\mu_0 > \bar{y} - t_{n-1}^{\alpha} s/\sqrt{n}. \tag{1.3}$$

However, one should use and report one-sided t tests and P values only when one is absolutely certain a priori of the direction of the difference if it is to occur.

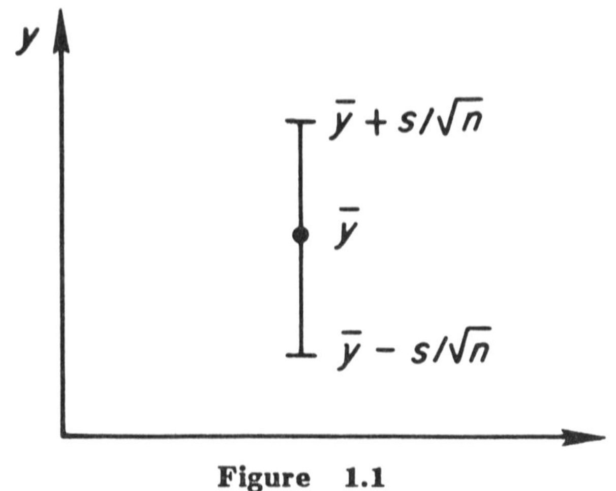

Figure 1.1

1.2. Nonnormality.

1.2.1. Effect

What happens when $F(y)$, the cdf of y, is not normal?* For large samples the t analysis is rescued by the central limit theorem:**

$$\sqrt{n}(\bar{y} - \mu_0) \xrightarrow{d} N(0, \sigma^2), \tag{1.4}$$

and also

$$s \xrightarrow{p} \sigma, \tag{1.5}$$

where $\sigma^2 = \text{Var}(y)$, so as $n \to \infty$,

$$t = \sqrt{n}(\bar{y} - \mu_0)/s \xrightarrow{d} N(0, 1). \tag{1.6}$$

Since $t_{n-1}^{\alpha} \to z^{\alpha}$ where z^{α} is the upper 100α percentile of the normal distribution, the t analysis will be valid in the limit.

---

* "cdf" denotes cumulative distribution function.

** "$\xrightarrow{d}$" and "$\xrightarrow{p}$" denote convergence in distribution and in probability, respectively.

How large is large? The answer to this is inextricably linked with how nonnormal F is. To discuss this it is necessary to introduce the two parameters that play a central role in the effects of nonnormality. They are the *skewness*

$$\gamma_1 = \gamma_1(y) = \frac{E(y-\mu)^3}{\sigma^3} \tag{1.7}$$

and the *kurtosis*

$$\gamma_2 = \gamma_2(y) = \frac{E(y-\mu)^4}{\sigma^4} - 3 \tag{1.8}$$

of the distribution.

For the normal distribution $\gamma_1 = \gamma_2 = 0$. For a distribution with a right tail heavier than its left γ_1 will be positive. As an example, the exponential distribution with $F'(y) = f(y) = \lambda\exp(-\lambda y)$, λ, $y > 0$, has $\gamma_1 = 2$. Similarly, for a distribution skewed to the left γ_1 will be negative. When the tails of the distribution contain more mass than the normal, the kurtosis γ_2 will be positive. For example, the two-tailed exponential (Laplace) distribution $f(y) = (\lambda/2)\exp(-\lambda|y|)$, $\lambda > 0$, $-\infty < y < +\infty$, has $\gamma_2 = 3$. The t distribution with ν df, which also has heavier tails than the normal, has $\gamma_2 = 6/(\nu-4)$ for $\nu > 4$, whereas the stubbier tailed uniform distribution has $\gamma_2 = -1.2$. For any distribution $\gamma_2 \geq -2$.

In the numerator of the t statistic (1.1)

$$E(\bar{y}) = \mu, \qquad \text{Var}(\bar{y}) = \frac{\sigma^2}{n} \tag{1.9}$$

for any distribution $F(y)$, and also

$$\gamma_1(\bar{y}) = \frac{\gamma_1}{\sqrt{n}}, \qquad \gamma_2(\bar{y}) = \frac{\gamma_2}{n}, \tag{1.10}$$

where $\gamma_1(\bar{y})$, $\gamma_2(\bar{y})$ are the skewness and kurtosis of the cdf of $\bar{y}$. From (1.10) one can infer that a kurtosis effect is wiped out rapidly whereas skewness vanishes more gradually. For most distributions the central limit theorem will have had time to weave its magic on $\bar{y}$ by $n = 10$, except possibly for a slightly skewed appearance.

In the denominator of the t statistic

$$E(s^2) = \sigma^2, \qquad \operatorname{Var}(s^2) = \sigma^4 \left(\frac{2}{n-1} + \frac{\gamma_2}{n} \right), \tag{1.11}$$

which transform approximately to

$$\begin{aligned} E(s) &\cong \sigma - \frac{\sigma}{8} \left(\frac{2}{n-1} + \frac{\gamma_2}{n} \right), \\ \operatorname{Var}(s) &\cong \frac{\sigma^2}{4} \left(\frac{2}{n-1} + \frac{\gamma_2}{n} \right). \end{aligned} \tag{1.12}$$

For $\gamma_2 > 0$ the convergence of s to σ will be slower than prophesied by the normal distribution whereas for $\gamma_2 < 0$ it will be faster.

Except in the case of the normal distribution the numerator and denominator are stochastically dependent. The asymptotic correlation between $\bar{y}$ and s is

$$\frac{\gamma_1}{\sqrt{\gamma_2 + 2}}, \tag{1.13}$$

which vanishes only if $\gamma_1 = 0$.

Power series expansions for the moments of t appear in the work of Geary (1936, 1947). The leading terms in the mean and variance are

$$\begin{aligned} E(t) &= -\frac{\gamma_1}{2\sqrt{n}} + 0\left(\frac{1}{n^{3/2}} \right), \\ \operatorname{Var}(t) &= 1 + \frac{1}{n}\left(2 + \frac{7}{4}\gamma_1^2 \right) + 0\left(\frac{1}{n^2} \right). \end{aligned} \tag{1.14}$$

This suggests that γ_2 has little effect on t but that γ_1 may have a larger effect.

The Monte Carlo sampling work of Pearson (1929) is in accord with the (later) moment calculations of Geary. Pearson considered different distributions with γ_1 ranging between 0 and .7 and γ_2 between $-.5$ and 4 and sample sizes $n = 2,\ 5,\ 10,\ 20$. For γ_1 and γ_2 in these ranges their effect on the distribution of $|t|$ is small. For $\gamma_2 > 0$ the actual two-sided P values tend to be smaller than the stated P values based on the t table, and for $\gamma_2 < 0$ the actual P values can be larger than the stated ones. Nonzero γ_1 tends to make the P values larger than the values calculated from the t tables. Later work by Gayen (1949) indicates that for values of γ_1, γ_2 outside these ranges (i.e., $\gamma_1 > 1$, $\gamma_2 > 4$) the robustness of t deteriorates rapidly for small n.

The situation is worse for one-sided P values based on t rather than $|t|$. The skewness of t has the leading term (from Geary, 1936, 1947)

$$\gamma_1(t) = -\frac{2\gamma_1}{\sqrt{n}} + 0\left(\frac{1}{n^{3/2}}\right). \tag{1.15}$$

The skewness is in the opposite direction from the parent population; this is caused by the correlation (1.13) between $\bar{y}$ and s. The tail probabilities in the skewed direction of t will be underestimated by the t table and overestimated in the opposite direction. These miscalculations cancel each other in obtaining two-sided P values, but for one-sided values the effect can be worrisome. As an illustration, Gayen (1949) showed that for $n = 10$ $P\{t < -2.262\}$ is .064 for a distribution with $\gamma_1 = \gamma_2 = 1$ rather than the nominally stated .025.

These calculations are confirmed in the Monte Carlo work of Pearson and Please (1975), who tabulated the fractions of samples falling above, below, and outside the appropriate $\alpha = .05$ and .01 t critical limits for various combinations of $n = 10, 20, 25$, $\gamma_1 = 0\,(.2).8$, and γ_2 in the range -1 to 1.4.

The special case where $y_i = u_i - v_i$ tends to be more robust. If the u and v distributions are identical except for location, or at least have approximately the same skewness $\gamma_1(u) \cong \gamma_1(v)$, then the differencing operation on $u - v$ will cancel out the skewness effect so that $\gamma_1(y) = 0$, or in the approximate case $\gamma_1(y) \cong 0$. The kurtosis

$\gamma_2(y)$ will most likely be nonzero, but since its effect on the P values is less than $\gamma_1(y)$, the t test should be more robust for this special case.

Efron (1969) studied extensively the behavior of the t statistic under the condition that the y_i are symmetrically distributed, which, of course, implies $\gamma_1(y) = 0$. His results suggest that under the symmetry assumption the t test often tends to be conservative; i.e., the true P values are less than the nominally stated ones. The effect is not large except for extreme distributions like the Cauchy.

Although this discussion points out that the user cannot go too far wrong with the t statistic, the reader should not come away with the impression that it is the best thing to use. For distributions other than the normal it is not the most efficient procedure and for some it can be very inefficient. Inefficiency means that the power of the test is not as great for alternative distributions as for other procedures more tailored to the underlying distributions. Correspondingly, the P values do not tend to be as impressively small when based on the t statistic as when they are derived from the specially designed tests. This means that whereas the t test is somewhat *robust for validity*, it is not *robust for efficiency*.

For example, if for a positive random variable it is quite clear from plotting the tail of the sample cdf (i.e., $1 - \hat{F}(y)$) on log paper (i.e., linear $\times$ logarithmic scales) that F is an exponential distribution, then the most powerful one-sided procedure uses $\bar{y}$ without s to compute a P value from the gamma distribution. If the data do not unequivocally demonstrate an exponential distribution but the distribution does have a long upper tail, then a transformation like log or square root (see Section 1.2.3) before the t statistic is computed will produce sharper results.

Another type of nonnormality that can occur is the appearance of *outliers*. These are observed values which are substantially remote from the main body of the data but cannot be discarded as being erroneous measurements, miscalculations, etc. They are judged not to have come from the distribution governing the rest of the data.

Whether outlying values are outliers or merely extreme observations from a heavy-tailed distribution is a fuzzy issue in many cases. Typically, aberrant values are considered outliers if they are few in number and the rest of the sample looks normally distributed with them removed. In Monte Carlo studies outliers are frequently modeled by having 95% of the sample come from a unit normal distribution and 5% from a normal distribution with $\mu = 0$ and $\sigma = 10$. Mixture distributions where with probability p the observation is distributed as $N(\mu, \sigma^2)$ and with probability $1 - p$ as $N(\mu, (k\sigma)^2)$ are referred to as *contaminated normal distributions*.

The effect of outliers on the sample mean can be noticeable, particularly if more occur in one tail than the other. However, the dramatic impact is on the sample variance. Because the differences from the mean are squared in the sample variance, squares from outliers can constitute a substantial fraction of the sum of squares even though they are few in number. The result is to inflate the denominator of the t statistic and consequently to dampen or wipe out an otherwise significant mean difference. Thus neither the mean nor the variance, especially the latter, is *resistant* to outliers.

An excellent treatise on outliers is Barnett and Lewis (1978).

1.2.2. Detection

My recommendation for detecting nonnormality is *probit plotting*. Probit plotting is facilitated by probit paper, which is specially constructed graph paper available from many companies under the name probability or normal probability paper. One scale is linear, and the other scale is designed to transform the cumulative normal distribution function into a straight line. A piece of probit paper resembles

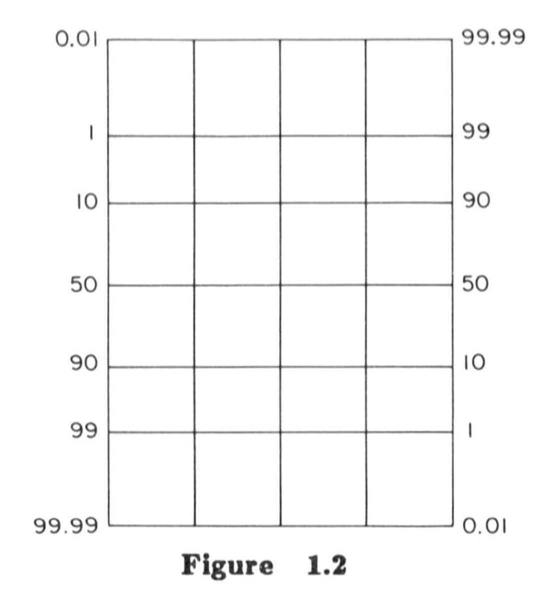

Figure 1.2

Figure 1.2 with many more lines for scale divisions. Since the cumulative normal distribution would require an infinite linear strip to reach 0 and 1 the probit scale is cut off, usually at .0001 and .9999. Note that most papers use a percent scale .01 to 99.99.

The paper is used in the following fashion. Form the ordered values $y_{(1)} \leq y_{(2)} \leq \cdots \leq y_{(n)}$ from the sample $y_1, \cdots, y_n$. Above the abscissa value $y_{(i)}$ on the linear scale plot a point at the ordinate value $i/(n+1)$. There is nothing sacred about the choice of $i/(n+1)$; another simple possibility is $(i - \frac{1}{2})/n$. The usual choice i/n for plotting the sample cdf is excluded because it goes off the scale at $i = n$. Chernoff and Lieberman (1954) have studied the optimal selection of the ordinate value from the point of view of estimating σ, but since the graph in this instance is merely intended for visual inspection of the tails of the distribution, the most computationally convenient choice suffices. On a computer it doesn't matter, but for hand plotting $i/(n + 1)$ is quite easy.

The points can be connected by straight lines if the plotter so desires, but this is not necessary. If a y value is repeated in the sample, then the sequence of points (or line) will proceed straight up at that value.

The abscissa value at which the sequence of points (or line segment) crosses the ordinate value 50% is the sample median, and in the case of the normal distribution this is an estimate of μ, though not the best one. The difference between the 84% and 50% points on the abscissa (and/or between the 50% and 16% points) is a quick estimate of σ for the normal distribution.

The observer is interested in how well the points $(y_{(i)}, i/(n+1))$, $i = 1, \cdots, n$, conform to a straight line. Deviation in the tails, not fluctuation in the middle, is what is important for inferences on μ. A sample like that depicted in Figure 1.3 is indicative of a distribution with $\gamma_1 > 0$. The more it bends at the top the shakier the t test gets, particularly one-sided P values. Figures 1.4 and 1.5 illustrate samples from distributions with $\gamma_2 > 0$ and $\gamma_2 < 0$, respectively.

Outliers give a slightly different appearance in probit plots, although the difference is unclear at times. Typically, the body of the data follows a straight line on probit paper, but there are a few values too far to the right (or left) as in Figure 1.6.

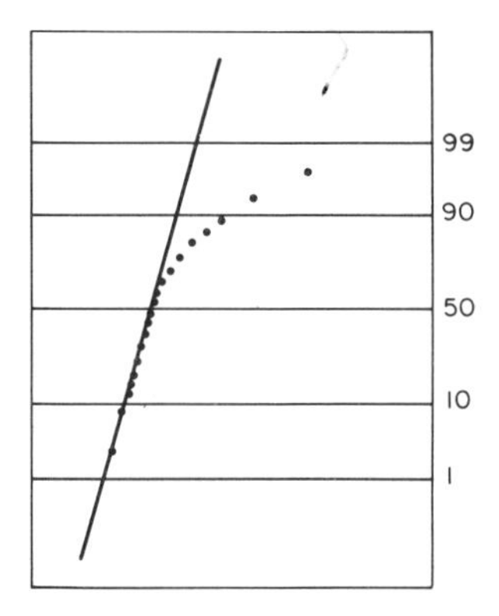

Figure 1.3

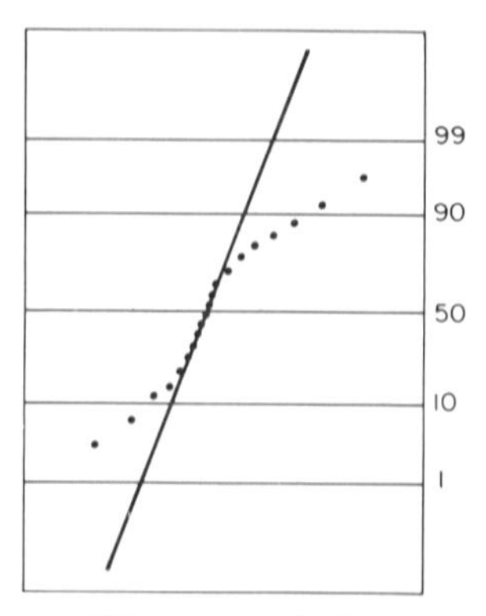

Figure 1.4

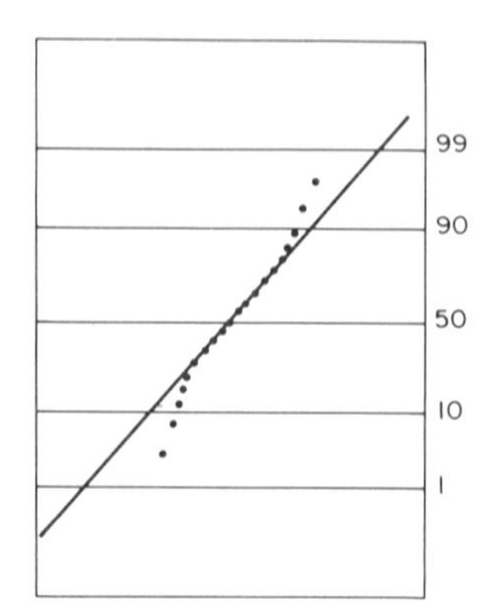

Figure 1.5

I make an effort to obtain a probit plot of the data before using the t test in any kind of crucial analysis. If the analysis requires many different t tests on different data sets, I at least try to plot some of the representative sets. Alternatively, one can ask the computer to do the plotting if it has a graphics routine for displaying $\Phi^{-1}(i/(n+1))$ versus $y_{(i)}$.*

The reader should be aware that log-probit paper exists as well. This has a normal probability scale on one axis and a logarithmic

* $\Phi(\cdot)$ is the cdf for $N(0,1)$.

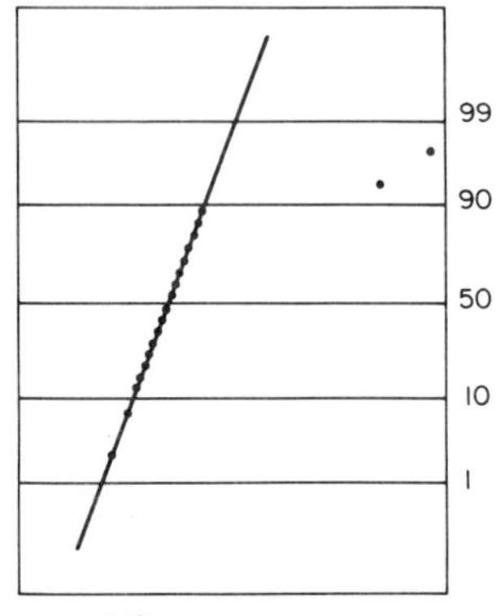

Figure 1.6

scale (1, 2, or 3 cycles for base 10) on the other. It is useful for examining whether the data are normally distributed after a logarithmic transformation.

Probit plotting is a special case of general quantile-quantile or *Q-Q plotting.* For further discussion see Wilk and Gnanadesikan (1968).

If a deviation from normality cannot be spotted by eye on probit paper, it is not worth worrying about. I never use the Kolmogorov-Smirnov test (or one of its cousins) or the χ^2 test as a preliminary test of normality. They do not tell you how the sample is differing from normality, and I have a feeling they are more likely to detect irregularities in the middle of the distribution than in the tails. If plotting is impractical for large data bases and some normality screening device is required, I would be inclined to compute either the sample estimates of γ_1 and γ_2 or the Shapiro-Francia test statistic, which are described next.

The sample estimates of γ_1 and γ_2 are

$$\hat{\gamma}_1 = \frac{1}{n}\sum_{i=1}^{n}(y_i - \bar{y})^3 \Big/ \left[\frac{1}{n}\sum_{i=1}^{n}(y_i - \bar{y})^2\right]^{3/2},$$
$$\hat{\gamma}_2 = \frac{\frac{1}{n}\sum_{i=1}^{n}(y_i - \bar{y})^4}{\left[\frac{1}{n}\sum_{i=1}^{n}(y_i - \bar{y})^2\right]^2} - 3. \tag{1.16}$$

These convey information about what type of departure from normality is occurring, and their values could be compared with the ranges (see Section 1.2.1) in which the t test is known to be robust. Mental allowance can be made for the sampling variability in these estimates. Tables of critical points for testing $\gamma_1 = 0$ or $\gamma_2 = 0$ appear in Pearson and Hartley (1970), but preliminary testing does not seem germane.

For testing normality Shapiro and Francia (1972) proposed the test statistic

$$W' = \frac{\left(\sum_{i=1}^{n} b_i y_{(i)}\right)^2}{\sum_{i=1}^{n}(y_i - \bar{y})^2}, \tag{1.17}$$

where $y_{(1)} \leq \cdots \leq y_{(n)}$ and

$$b_i = \frac{m_i}{\left(\sum_{i=1}^{n} m_i^2\right)^{1/2}}, \qquad m_i = E(z_{(i)}) \tag{1.18}$$

with $z_{(1)} \leq \cdots \leq z_{(n)}$ representing the order statistics from a unit normal distribution. The idea behind the statistic (1.17) is that if the y_i are normally distributed, then the correlation between the $y_{(i)}$ and their expected values under normal theory should be very high. Rejection of normality should be for low values of W'.

Since the correlation coefficient is location and scale invariant, the expected values can be taken to be those for order statistics from a unit normal distribution. Tables of m_i are available in Harter (1961) for $n = 2(1)100(25)300(50)400$; values for additional $n > 100$ can be found in Harter (1969b). A small table of critical values for

W' is given by Shapiro and Francia.

The statistic W' is a simplification of a statistic W proposed earlier by Shapiro and Wilk (1965). Since the $y_{(i)}$ are not independent, Shapiro and Wilk take their covariance structure into account in the statistic

$$W = \frac{\left(\sum_{i=1}^{n} a_i y_{(i)}\right)^2}{\sum_{i=1}^{n}(y_i - \bar{y})^2}, \tag{1.19}$$

where

$$\begin{aligned} \mathbf{a}^T &= (a_1, \cdots, a_n) = \frac{\mathbf{m}^T \mathbf{V}^{-1}}{(\mathbf{m}^T \mathbf{V}^{-2} \mathbf{m})^{1/2}}, \\ \mathbf{m}^T &= (m_1, \cdots, m_n), \end{aligned} \tag{1.20}$$

and $\mathbf{V}$ is the covariance matrix of $(y_{(1)}, \cdots, y_{(n)})$. Currently available tables of $\mathbf{a}$ for W (see Shapiro and Wilk, 1965) are not nearly as extensive as those for $\mathbf{m}$ cited previously.

Shapiro, Wilk, and Chen (1968) have shown the Shapiro-Wilk test to be the best currently available procedure for testing normality.

There are also tests especially designed for detecting outliers (see Barnett and Lewis, 1978, Chapters 2 and 3, and Miller, 1981, Chapter 6). However, I am inclined to use only a procedure resistant to outliers (see Sections 1.2.3, "Nonparametric Techniques" and "Robust Estimation") if there is any possibility of their presence rather than to run a preliminary test.

1.2.3. Correction

Transformations One method of handling data that are sufficiently nonnormal to be worrisome is to seek a transformation that will convert the data into a sample that looks approximately normally distributed. With positive data, if they are not approximately symmetrically distributed, they are practically always positively skewed. For this circumstance the most commonly employed transformations are the *logarithmic transformation* $z = \log y$ (to the base 10 or e)

and the *square root transformation* $z = \sqrt{y}$. These are special cases of the power family

$$z = \begin{cases} \frac{y^\lambda - 1}{\lambda}, & \lambda \neq 0, \\ \log y, & \lambda = 0. \end{cases} \tag{1.21}$$

In practice, one would simply compute $z = y^\lambda$ when $\lambda \neq 0$, but the representation (1.21) shows how $\log y$ fits into the family. The log and square root transformations are more frequently used than other members of the power family because tables for them are readily available and many electronic calculators now have these routines programmed into the hardware so that the mere touch of a key will produce the transformed value. Of course, in large computers any member of the family is equally good.

Power transformations are mainly used only on positive random variables. The family can be generalized to

$$z = \begin{cases} \frac{(y+c)^\lambda - 1}{\lambda}, & \lambda \neq 0, \\ \log(y + c), & \lambda = 0, \end{cases} \tag{1.22}$$

which may be useful in instances where there is a finite negative lower bound to the possible value of the variable. However, for variables assuming positive and negative values it is more customary to use nonparametric methods, which will be described shortly. Addition (or subtraction) of a small constant may also improve the normality of the transformed values even for strictly positive variates, particularly those that can take values close to zero.

There are other special purpose transforms useful in data analysis like $\sin^{-1}\sqrt{\hat{p}}$ for the binomial estimator and $\tanh^{-1} r$ for the sample correlation coefficient from a bivariate normal distribution. These are designed to make the variance of the estimator relatively free of the unknown parameter, and at the same time they seem to improve the normal approximation. However, they are not particularly pertinent to the current discussion.

Selection of the appropriate transformation depends mostly on guesswork and experience. There has been theoretical work done to systematize the search for the best transform, and three notable articles in this direction are Tukey (1957), Box and Cox (1964), and Hinkley (1975). However, I would say that at the present day the most common practice is to let experience suggest a transform and then to check via a probit plot whether the guess is reasonably succesful. When there are two or more samples there is an empirical method for selecting a variance stabilizing transformation. Since stable variances and normality frequently seem to walk hand in hand, this method offers a substitute for guesswork in the multisample problem, discussed in Chapters 2 and 3.

For hypothesis testing the null hypothesis $H_0 : E(y) = \mu_0$ transforms under $z = g(y)$ into $H_0 : E(z) = g(\mu_0)$. Those of an exact mathematical mind will shudder at such crudity, but the correspondence is sufficient for practical purposes. Moreover, if z is more normally distributed than y, the transformed hypothesis $H_0 : E(z) = g(\mu_0)$ is probably a better statement of the null situation than the original null hypothesis. As an illustration, if the basic variable is a ratio $y = u/v$, then the log transform $z = \log y$ sometimes produces more Gaussian looking data, in which case the null hypothesis $H_0 : E(y) = 1$ transforms to $H_0 : E(z) = 0$, i.e., $E(\log u) = E(\log v)$.

If the null hypothesis is stated in terms of medians, then it transforms exactly under monotone transformations. That is, H_0 : median $y = \mu_0$ is precisely equivalent to H_0 : median $z = g(\mu_0)$ for $z = g(y)$, g monotone.

Transformations seldom are helpful in trying to handle outliers. An outlier typically remains an outlier after the square root or logarithmic transformations. Transformations strong enough to pull outliers into proximity with the rest of the data compress the data too much. Better avenues for handling outliers are through nonparametric methods or robust estimators.

Nonparametric Techniques An alternative approach for handling nonnormality is to use a nonparametric test statistic in place of the t statistic. There are many possible nonparametric tests, but I will mention only the three I consider most useful.

The first and simplest is the *sign test*. Initially let me asume the underlying cdf is continuous in order to avoid ties. The null hypothesis is that the median η of the distribution equals a specified value η_0; i.e., $H_0 : P\{y < \eta_0\} = P\{y > \eta_0\} = \frac{1}{2}$. No assumption of normality or even symmetry about η_0 is needed in the underlying model. The test statistic is

$$S = \sum_{i=1}^{n} I\{y_i > \eta_0\}, \tag{1.23}$$

where

$$I\{y_i > \eta_0\} = \begin{cases} 1 & \text{if } y_i > \eta_0, \\ 0 & \text{if } y_i < \eta_0; \end{cases} \tag{1.24}$$

i.e., S is the number of y_i which exceed η_0.

Under H_0 the statistic S has a binomial distribution with parameters n and $p = \frac{1}{2}$. The lower one-tailed P value is

$$P = \sum_{k=0}^{S} \binom{n}{k} \left(\frac{1}{2}\right)^n, \tag{1.25}$$

and this can easily be obtained from binomial tables (e.g., Harvard, 1955, or Owen, 1962). An analogous expression holds for an upper one-tailed P value, and the two-sided P value for $S < \frac{n}{2}$

$$P = \sum_{k=0}^{S} \binom{n}{k} \left(\frac{1}{2}\right)^n + \sum_{k=n-S}^{n} \binom{n}{k} \left(\frac{1}{2}\right)^n \tag{1.26}$$

also can be extracted easily from tables. For $S > \frac{n}{2}$ transpose S and $n - S$ in (1.26). For large n (viz., $n > 25$) the normal approximation

$$\frac{S - \frac{n}{2}}{\frac{1}{2}\sqrt{n}} \approx N(0, 1) \tag{1.27}$$

gives quite accurate P values.* Even for n as small as 10 I don't hesitate to resort to the approximation (1.27) if tables are not available. For upper tail P values subtraction (addition for lower tail) of $\frac{1}{2}$ in the numerator as a continuity correction will refine the approximation.

Whereas the t test is associated with the estimator $\bar{y}$ for the location of the population, the sign test is related to the median m of the sample. Confidence intervals for the population median can be determined from (1.26) or (1.27) by figuring out the range of η for which P is greater than α. If $s^{\alpha/2}$ is the critical value for S, i.e., the largest integer such that

$$\sum_{k=0}^{s^{\alpha/2}} \binom{n}{k} \left(\frac{1}{2}\right)^n + \sum_{k=n-s^{\alpha/2}}^{n} \binom{n}{k} \left(\frac{1}{2}\right)^n \leq \alpha \tag{1.28}$$

or

$$\Phi\left(\frac{s^{\alpha/2} - \frac{n}{2} + \frac{1}{2}}{\frac{\sqrt{n}}{2}}\right) + \left[1 - \Phi\left(\frac{n - s^{\alpha/2} - \frac{n}{2} - \frac{1}{2}}{\frac{\sqrt{n}}{2}}\right)\right] \leq \alpha, \tag{1.29}$$

then $(y_{(s^{\alpha/2}+1)}, y_{(n-s^{\alpha/2}-1)})$ is the ($\geq 100(1-\alpha)\%$) confidence interval for the population median η, where $y_{(1)} \leq \cdots \leq y_{(n)}$ are the order statistics. Tables are available in Owen (1962).

---

* "$\approx$" denotes "is approximately distributed as."

The sign test is not very efficient for many distributions in comparison with the t test or the signed-rank test, which is the next test to be discussed. It often throws away too much information, although for some very heavy-tailed distributions it is well to ignore the data except for their signs. For instance, the sign test is asymptotically optimal for the two-tailed exponential distribution, and it is better than the t test or signed-rank test for the Cauchy distribution. The sign test very effectively obliterates the effect of outliers.

I tend to use the sign test as a quick test or a screening device. If the data are clearly statistically significant and the sign test will prove this, it is a marvelous device for hurriedly getting the client out of your office. He or she will be happy because the data have received an official stamp of statistical significance, and you will be happy because you can get back to your own research. It is also useful for rapidly scanning data to acquire a feeling as to whether the data might be statistically significant. If the sign statistic and approximation (1.27) produce a normal deviate which is near to being significant, then a more refined analysis may be worthwhile. If, on the other hand, S is nowhere close to being significant, it is very unlikely that a significant result can be produced by more elaborate means.

Until now I have kept the question of ties locked in the closet, but unfortunately they can, and do, occur. For calculating a P value the only ties that cause trouble are those in which y equals the null median η_0. For confidence intervals other ties can cause problems, but the reader is left to extrapolate the null discussion to the broader case.

If the possible values of u and v are discrete and relatively few, then in the paired data problem where $y = u - v$ a number of the observations may equal the null median 0. The conditional approach is to exclude the zeros and to consider the question

$P\{y > 0|y \neq 0\} \gtrless P\{y < 0|y \neq 0\}$. But this may be a worthless question to answer if $y = 0$ a high proportion of the time. If u and v are frequently identical, it may be unimportant which is selected, and other considerations such as cost or side-effects in medical applications may be more influential in the selection process. A significance level attached to the conditional data may be misinterpreted by the unwary.

For a small proportion of ties the conditional approach is easy and acceptable. A conservative stance would be to consider the zeros as having small values in the direction opposite to the shift of the rest of the data. If the sign test still gives a delightfully small P value, then one is quite content about the ties, but one may not be so lucky. A less self-penalizing procedure is to score each zero as one-half in calculating S. At no time would I use randomization to break the zero ties.

For a study of handling ties in nonparametric tests the reader is referred to Putter (1955).

When the analysis requires more than the sign test, my favorite is the *Wilcoxon* (1945) *signed-rank test*. The null hypothesis is that the underlying cdf is symmetric about a specified value μ_0, usually zero. Symmetry about μ_0 is used in the test procedure so a falsely significant result can be produced by asymmetry even though the mean or median equals μ_0.* To avoid ties at the outset assume the underlying cdf is continuous.

Subtract the hypothesized mean from each observation; i.e., let $z_i = y_i - \mu_0$. Take the absolute values $|z_1|, \cdots, |z_n|$ and order them $|z|_{(1)} \leq \cdots \leq |z|_{(n)}$. Identify with each absolute value its rank from 1 up to n. For z_i let r_i be the rank of its absolute value. The Wilcoxon signed-rank statistic is the sum of the ranks corresponding to positive

---

* The test is consistent against alternatives for which $P\{y_1 + y_2 > 0\} \neq 1/2$.

observations, i.e.,

$$SR_+ = \sum_{i=1}^{n} r_i \, I\{z_i > 0\}, \tag{1.30}$$

where

$$I\{z_i > 0\} = \begin{cases} 1 & \text{if } z_i > 0, \\ 0 & \text{if } z_i < 0. \end{cases} \tag{1.31}$$

Since the sum of all the ranks equals $n(n+1)/2$, equivalent statistics are the sum of the negative ranks or the difference between the positive and the negative ranks.

An alternative representation for the Wilcoxon signed-rank statistic, which the reader can verify with a little thought or mathematical induction, is

$$SR_+ = \sum_{i=1}^{n} \sum_{j=1}^{i} I\{z_i + z_j > 0\}, \tag{1.32}$$

where

$$I\{z_i + z_j > 0\} = \begin{cases} 1 & \text{if } z_i + z_j > 0, \\ 0 & \text{if } z_i + z_j < 0. \end{cases} \tag{1.33}$$

In most instances (1.30) is the easier way to compute SR_+, but (1.32) is theoretically convenient for computing moments and studying distribution theory. The representation (1.32) is due to Tukey.

The probabilities $P\{SR_+ = r\}$ can be generated through recursive schemes, and tables are readily available. Two compendia containing signed-rank tables are Owen (1962) and Pearson and Hartley (1972). They give cumulative probabilities for values of n up to 20 and 15, respectively. Beyond this the normal approximation

$$\frac{SR_+ - \frac{n(n+1)}{4}}{\sqrt{\frac{n(n+1)(2n+1)}{24}}} \approx N(0,1) \tag{1.34}$$

is sufficient for computing one- or two-tailed P values.

The estimator associated with the signed-rank statistic is the *Hodges-Lehmann* (1963) *estimator*, which is the median of the $n(n+1)/2$ values $(y_i + y_j)/2$ where i can equal j. The connection is suggested by the representation (1.32). This leads into the field of robust estimators for symmetric distributions, which is discussed next in Section 1.2.3.

Although it is not found frequently, a confidence interval for the population median can be constructed from the signed-rank statistic. The ($\geq 100(1-\alpha)\%$) interval consists of all values of μ such that when SR_+ is computed for $z_i = y_i - \mu$, $i = 1, \cdots, n$, the two-sided P value is greater than or equal to α. It is a bit tedious to figure out the interval through guesswork or trial and error, which is probably the reason for its lack of popularity. However, there is a graphical procedure due to Tukey which greatly simplifies this process. On a piece of graph paper plot the n points $y_1, \cdots, y_n$ on the ordinate axis. Through each point y_i draw two lines in the right half-plane, one with slope $+1$, the other with slope -1. These lines will intersect at $\binom{n}{2}$ points in the right half-plane. These intersections and the original n points give a total of $n(n+1)/2$ points whose ordinates constitute the collection $\{(y_i + y_j)/2\}$. The median of these ordinal values is the Hodges-Lehmann estimator. If $sr_+^{\alpha/2}$ is the critical value for SR_+ (i.e., the largest integer such that $P\{SR_+ \leq sr_+^{\alpha/2}$ and $SR_+ \geq n - sr_+^{\alpha/2} \mid H_0\} \leq \alpha$), then the $sr_+^{\alpha/2} + 1$ smallest ordinate in the collection is the lower confidence limit and the $n(n+1)/2 - sr_+^{\alpha/2}$ largest (i.e., $sr_+^{\alpha/2} + 1$ from the top) ordinate is the upper limit. For $n = 5$, $sr_+^{\alpha/2} = 3$ the procedure is illustrated in Figure 1.7. Note that the Hodges-Lehmann estimator need not be the midpoint of the confidence interval.

What should be done about ties? For the signed-rank test ties between values of $z = y - \mu_0$ with the same absolute values but opposite signs causes problems as well as those for which $z = 0$. The

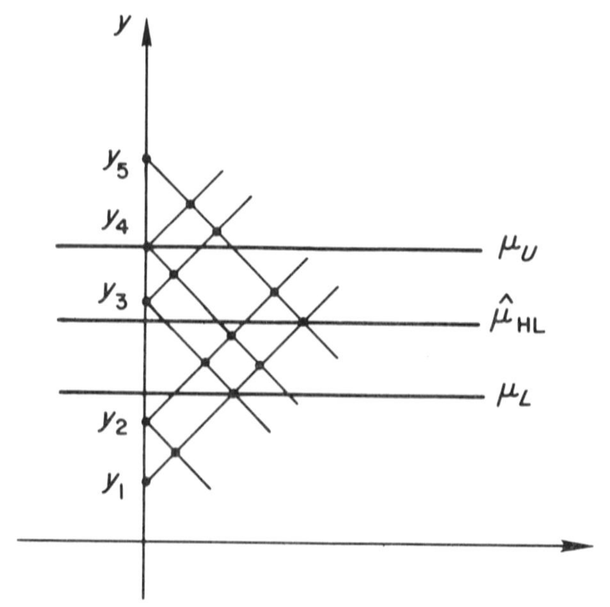

Figure 1.7

zeros can be dropped and the test performed conditionally without them as in the case of the sign test. Pratt (1959) has pointed out that anomalies can occur with this approach but the circumstances seem rare. The more major question is whether it is worth investigating any shift of the conditional distribution if the probability of a zero value is large. For nonzero ties the successive ranks can be averaged and the average rank assigned to each observation in the tie. This is equivalent to expanding the definition (1.33) to

$$I\{z_i + z_j > 0\} = \begin{cases} 1 & \text{if } z_i + z_j > 0, \\ \frac{1}{2} & \text{if } z_i + z_j = 0, \\ 0 & \text{if } z_i + z_j < 0. \end{cases} \tag{1.35}$$

For a small number of average ranks the usual tables can be used with impunity. The variance of SR_+ corrected for ties is

$$\frac{1}{24}\left[n(n+1)(2n+1)-\frac{1}{2}\sum_{k=1}^{g}t_k(t_k-1)(t_k+1)\right], \tag{1.36}$$

where g is the number of tied groups and t_k is the size of the kth group. The square root of (1.36) can be substituted into the denominator of (1.34). However, the number of ties has to become considerable before the correction term in (1.36) makes much difference.

Pratt (1959) has the most thorough discussion of ties for the signed-rank test, and it is an article worth reading. He proposes a modified procedure for handling zero ties which deletes the ranks assigned to zeros. Cureton (1967) gives the null mean and variance for Pratt's statistic, and Rahe (1974) provides small sample tables. Conover (1972) gives some theoretical efficiencies for the different procedures.

As with the sign test, the signed-rank test is good for handling heavy-tailed distributions and outliers. Also, it is asymptotically optimal for the logistic distribution. Its asymptotic relative efficiency with respect to the t test for the normal distribution is $\frac{3}{\pi}$.

There are other nonparametric tests which, like the signed and signed-rank tests, sum a set of scores for the positive observations. An important example is the normal scores test (see Lehmann, 1975, pp. 96–97). This test requires specialized tables even for the computation of the statistic and therefore is inconvenient to use, even on a large computer. Also, the normal scores test outperforms the signed-rank test for short-tailed distributions like the uniform, but these are not as much of a worry as the heavy-tailed distributions where the signed-rank does better (see Hodges and Lehmann, 1961). Of the class of linear rank tests the sign and the signed-rank tests are by far the most important for applications.

The last nonparametric test to be mentioned is of a different type. It is the *Fisher* (1935) *permutation test.* The null hypothesis is that the underlying cdf is symmetric about μ_0. The test is linked with the estimator $\bar{y}$ since it uses $T = \sum_{i=1}^{n}(y_i - \mu_0)$ as a test statistic. Under H_0 the values $\pm(y_i - \mu_0)$ are equally likely so the 2^n different values of T with all possible sign changes $\sum_{i=1}^{n} \pm(y_i - \mu_0)$ are equally likely.* These T values can be ordered $T_{(1)} \leq \cdots \leq T_{(2^n)}$ and the one-sided P value equals the number of T values equal to or more extreme than the observed in that tail divided by 2^n. The two-sided P value equals the number of T values equal to or more extreme than the observed in both tails divided by 2^n.

The test is clumsy to carry out unless the observed T value is so large positively (or negatively) that only a few easily recognizable cases exceed it. For this reason it is seldom used. However, the idea behind the test can be extremely useful in situations more complicated than the one sample problem. In a complex model the statistician may be able to construct a score function which should be sensitive to detecting the type of alternatives suspected. Under the null hypothesis it will usually be random as to which group an observation belongs so the computer can generate all possible values of the score function that will be equally likely under randomization theory. If the total number of permutations is too large even for the computer, the computer can at least generate a large number of random permutations which will give an estimated P value.

There is no reason the permutation test has to use the statistic $\bar{y} - \mu_0$. It could just as well use the trimmed mean, which is to be mentioned shortly. If the regular mean difference $\bar{y} - \mu_0$ were divided by $s/\sqrt{n}$ to give the t statistic, the ordering of the values would be undisturbed because the term $\sum_{i=1}^{n}(y_i - \mu_0)^2$ in $\sum_{i=1}^{n}(y_i - \bar{y})^2 =$

---

* If there are k values of y_i which equal μ_0, then the problem reduces to considering 2^{n-k} different possible values of T.

$\sum_{i=1}^{n}(y_i - \mu_0)^2 - n(\bar{y} - \mu_0)^2$ is constant under sign changes. This would not be true if the sample was first trimmed so it may be best to employ a standardized statistic if using the trimmed mean.

The permutation technique can be used to construct a confidence interval for μ by calculating the range of μ_0 values which fail to give a P value less than or equal to α. The computations are usually too cumbersome, however, unless n and α are quite small.

Ties cause no problems for the permutation test. Zeros are treated in a conditional fashion as though the sample were smaller and had no zeros although one could replace them by a small number to see what effect breaking the ties might have on the P value.

The permutation test does not reduce the effect of large observations as the sign and signed-rank tests do. Not surprisingly, it is asymptotically equivalent to the t test. However, for small samples it can give more robust P values than the t ratio.

A variation of the permutation idea is to sample with replacement from the observed values. This is called the *bootstrap method*. For details see Efron (1979, 1982).

Robust Estimation The field of robust estimation for the location of a symmetric distribution has undergone intense investigation since the late 1960's. Major works that will permit the reader to enter the literature of this field are Andrews et al. (1972) and Huber (1977, 1981).

The three principal categories of robust estimators are the L, M, and R-estimators. An *L-estimator* is a linear combination of order statistics. The median, the mean, and the trimmed mean are the most important examples of L-estimators. An *M-estimator* is the root of the equation

$$\sum_{i=1}^{n} \psi((y_i - \theta)/S) = 0, \tag{1.37}$$

where the ψ function and scale estimate S are selected by the statistician. Some maximum likelihood estimators, like $\bar{y}$ for the normal distribution, are special cases since the derivative of the log likelihood (i.e., $f'(x)/f(x)$) is a ψ function. The median is also an M-estimator. Finally, *R-estimators* are linked to rank tests. The primary example of an R-estimator is the Hodges-Lehmann estimator.

Besides the median the most important robust estimator for applications is the *trimmed mean*. Let δ be some small proportion such as .10 or .05. The trimmed mean $\bar{y}_T$ discards δn (assumed here to be an integer) observations from each tail and computes the mean of the remaining observations.* If $y_{(1)} \leq \cdots \leq y_{(n)}$, then

$$\bar{y}_T = \frac{1}{(1-2\delta)n} \sum_{i=\delta n+1}^{n-\delta n} y_{(i)}. \tag{1.38}$$

The trimmed mean eliminates the effect of tail observations, be they from a heavy-tailed distribution or outliers. However, unless the trimming is used to remove really aberrant values, I have frequently found that the change from the mean has been only slight and is of little interest to the investigator. See Stigler (1977) for comparisons on real data.

The appropriate variance for use with the trimmed mean is the winsorized variance. Generally, winsorization (named after C. P. Winsor) replaces tail order statistics by a smaller (larger) order

---

* This technique is used in judging diving competitions where the highest and lowest scores from the judges are discarded before computing the average score for the dive. This average score is then multiplied by the degree of difficulty of the dive.

statistic. Specifically, let

$$y_{W(i)} = \begin{cases} y_{\delta n+1}, & i = 1, \cdots, \delta n, \\ y_{(i)}, & i = \delta n + 1, \cdots, n - \delta n, \\ y_{(n-\delta n)}, & i = n - \delta n + 1, \cdots, n, \end{cases} \tag{1.39}$$

where the fraction δ is the same as for the trimmed mean. Then

$$\begin{aligned} \bar{y}_W &= \frac{1}{n} \sum_{i=1}^{n} y_{W(i)}, \\ &= \frac{1}{n} \left[\delta n \, y_{(\delta n+1)} + \sum_{i=\delta n+1}^{n-\delta n} y_{(i)} + \delta n \, y_{(n-\delta n)} \right] \end{aligned} \tag{1.40}$$

is the *winsorized mean*, and the *winsorized variance* is*

$$\begin{aligned} s_W^2 &= \frac{1}{(1-2\delta)^2(n-1)} \sum_{i=1}^{n} (y_{W(i)} - \bar{y}_W)^2, \\ &= \frac{1}{(1-2\delta)^2(n-1)} \Big[\delta n (y_{(\delta n+1)} - \bar{y}_W)^2 \\ &\quad + \sum_{i=\delta n+1}^{n-\delta n} (y_{(i)} - \bar{y}_W)^2 + \delta n (y_{(n-\delta n)} - \bar{y}_W)^2 \Big]. \end{aligned} \tag{1.41}$$

For symmetric distributions a consistent estimate of the asymptotic variance of the trimmed mean is s_W^2/n, i.e.,

$$\widehat{\text{AVar}}(\bar{y}_T) = \frac{1}{n} s_W^2. \tag{1.42}$$

This is most easily established through the influence function (see Hampel, 1974).

---

* Some authors use $(1-2\delta)^2 n$ or $(1-2\delta)[(1-2\delta)n-1]$ for the denominator of s_W^2. Expression (1.41) allows standard programs for the variance to be applied to the winsorized sample; the calculated variance is then corrected by the factor $(1-2\delta)^{-2}$.

Asymptotically valid tests and confidence intervals for the mean μ of a symmetric distribution can be constructed from the relation

$$\frac{\sqrt{n}(\bar{y}_T - \mu)}{s_W} \xrightarrow{d} N(0,1). \tag{1.43}$$

For small sample sizes one might want to use a t interval such as

$$\bar{y}_T - t_\nu^{\alpha/2} \frac{s_W}{\sqrt{n}} < \mu < \bar{y}_T + t_\nu^{\alpha/2} \frac{s_W}{\sqrt{n}}. \tag{1.44}$$

Tukey and McLaughlin (1963) suggested that the degrees of freedom be taken to be $\nu = n(1-2\delta) - 1$, i.e., one less than the number of observations entering the trimmed mean. Monte Carlo work by Gross (1976) for $n = 10$ and 20 substantiates that this is approximately correct for normal distributions. Further substantiation can be found in the Monte Carlo study of Yuen and Dixon (1973) on the two sample problem. For a variety of heavy-tailed distributions Gross also found that the intervals (1.44) with the suggested degrees of freedom are conservative; that is, the true coverage is higher than the nominally stated coverage. For example, for the Cauchy distribution the true coverage is 97.5% when $n = 10$ and 97% when $n = 20$ with $\alpha = .05$ and $\delta = .10$.*

The class of M-estimators has received a great deal of theoretical attention, but M-estimators are not standardly used in practice at this time, although this may be changing. A prominant M-estimator is the *Tukey bisquare* (or *biweight*) *estimator*, which uses the ψ function

$$\psi(x) = \begin{cases} x(1-x^2)^2, & |x| \le 1, \\ 0, & |x| > 1, \end{cases} \tag{1.45}$$

---

* For these calculations, Gross (1976) used n instead of $n-1$ in the denominator of (1.41) and his critical constant $t^*_{\max}$ instead of $t_\nu^{\alpha/2}$.

and the scale estimate

$$\begin{aligned} S &= k \cdot MAD, \\ &= k \cdot \text{median}\{|y_i - m|, \quad i = 1, \cdots, n\}, \end{aligned} \tag{1.46}$$

where m is the sample median and the arbitrary constant k is commonly 7.4, 8.2, or 9.0 (see Andrews et al., 1972). This estimator has good efficiency for the normal distribution and a variety of heavy-tailed distributions. Other frontrunners are the sine wave estimator of Andrews, the redescending linear segment estimator of Hampel, and the nonredescending linear segment estimator of Huber. Gross (1976) studies the confidence interval procedures associated with each of these estimators, with the exception of the last one of Huber.

The aforementioned robust estimators are predicated on the assumption that the underlying distribution is symmetric about its median. Symmetry is fundamentally used in the estimators and their variance estimators. What does one do if the empirical distribution appears asymmetric? No corresponding body of theory of robust estimators exists for asymmetric distributions at the present time. Some hardy souls recommend continued use of symmetric robust estimators on the grounds that it is difficult to tell from the sample whether the true underlying distribution is symmetric, but I cannot recommend this. I would be more likely to seek a transformation that symmetrizes the body of the data and then apply a robust estimator to the transformed data.

1.3. Dependence.

Although anything is possible, there are mainly just two types of dependence which arise in the applications envisaged in this book. Often the scientific investigator may be unaware of the importance to the statistical analysis of factors that can cause these dependencies so it is the responsibility of the statistician to ferret out by cross-examination of the investigator and/or examination of the data whether any effect exists.

The first type of dependence is caused by a *blocking* effect. The n data points $y_1, \cdots, y_n$ may have been collected in subgroups. For instance, some y may come from experiments on one day, others from different days. Or some y may be observations on animals in the same cage or litter whereas other y come from different cages or litters. The investigator usually will be cognizant of factors built into the experiment such as days, lab technicians, or litters, but may not be careful about informing the statistician of the presence of these nuisance factors.

Maybe a nuisance factor has no effect, but one should not just asume this. For unbalanced blocking the estimates can be biased, and the error variance is always distorted. The standard way of detecting and correcting for block effects is to remodel the problem into a higher-way classification with fixed and random effects. Since this solution is fairly universally understood and covered to some extent in later chapters of this book, it is not discussed in detail now.

The other type of dependence can come from a *sequence* effect. The sequence may be in time or space. The observations may be taken serially in time in which case observations close together in time may be stochastically dependent due to slow random variations in the experimental conditions or instrumentation, or due to an observation having a direct effect on the next succeeding observation. Similarly, observations on objects located physically next to each other may be dependent through greater similarity of local conditions or through direct interaction between the objects.

We shall examine the simplest possible sequence effect where there is a *serial correlation* of lag 1. That is, for $i = 1, \cdots, n$,

$$\begin{aligned} y_i &\sim N(\mu, \sigma^2), \\ \text{Cov}(y_i, y_{i+1}) &= \rho_1 \sigma^2, \qquad (1.47) \\ \text{Cov}(y_i, y_{i+j}) &= 0, \quad j \neq 0, 1. \end{aligned}$$

The dependence could of course extend to lags greater than 1 (i.e., $\text{Cov}(y_i, y_{i+j}) = \rho_j \sigma^2$, $j > 1$), but this simplest case is an important one for data analysis and will illustrate the difficulties. In some problems the serial correlations $\rho_2, \rho_3, \cdots$ may be nonzero but appreciably smaller than ρ_1 in magnitude and thus not affect the analysis as much as ρ_1. However, for general serial dependence one is forced into time series analysis, which is beyond the scope of this book.

1.3.1. Effect

One can readily compute

$$\begin{aligned} E(\bar{y}) = \mu, \qquad \text{Var}(\bar{y}) &= \frac{\sigma^2}{n}\left[1 + 2\rho_1\left(1 - \frac{1}{n}\right)\right], \\ E(s^2) &= \sigma^2\left(1 - \frac{2\rho_1}{n}\right), \end{aligned} \qquad (1.48)$$

and show that $\text{Var}(s^2) \to 0$ as $n \to \infty$. Since $\bar{y}$ is normally distributed, this establishes that

$$\frac{\sqrt{n}(\bar{y} - \mu)}{s} \xrightarrow{d} N(0,\ 1 + 2\rho_1). \qquad (1.49)$$

The convergence (1.49) still holds even if the y_i are not normally distributed by the central limit theorem for m-dependent random variables (see Fraser, 1957, p. 219) so long as y_i and y_{i+j} are independent for $j > 1$.

The limiting variance $1+2\rho_1$ can be substantially different from 1 even for moderate values of ρ_1. This will produce discrepancies in the P value. For instance, if $\rho_1 = \frac{1}{3}$ the limiting standard deviation is 1.29 instead of 1 so for a t value equal to 1.96 the actual two-sided P value is .13 whereas the investigator unaware of ρ_1 would state $P = .05$. Clearly, the effect of ρ_1 on the P value can be most unpleasant.

Gastwirth and Rubin (1975) study the effects of serial dependence on robust estimators.

1.3.2. Detection

The methods of detection are the same as for examining the association between any pair of variables, which in this case are y_i and y_{i+1}. One can plot the pairs (y_i, y_{i+1}), $i = 1, \cdots, n-1$, and/or compute the sample serial correlation coefficient

$$r_1 = \frac{\frac{1}{n-1}\sum_{i=1}^{n-1}(y_i - \bar{y})(y_{i+1} - \bar{y})}{\frac{1}{n}\sum_{i=1}^{n}(y_i - \bar{y})^2}. \tag{1.50}$$

It is the size of r_1 that is important and not whether it is statistically different from zero. Thus a preliminary test of $\rho_1 = 0$ has little value, but for those so inclined a good reference is T. W. Anderson (1971).

The distribution theory for serial correlation coefficients is very difficult. Tables of critical values for the circular serial correlation coefficient are available in R. L. Anderson (1942), Dixon (1944), and T. W. Anderson (1971, p. 319). Under the null hypothesis and normal theory the circular serial correlation coefficient

$$r_1^* = \frac{\sum_{i=1}^{n}(y_i - \bar{y})(y_{i+1} - \bar{y})}{\sum_{i=1}^{n}(y_i - \bar{y})^2}, \tag{1.51}$$

where $y_{n+1} = y_1$ by definition, is approximately distributed as $r - (1/n)$ where r is the ordinary Pearson product-moment correlation coefficient based on $n + 3$ observations. This approximation is satisfactory for $n \geq 10$ and is very good for $n \geq 25$. For details see Hannan (1960, pp. 85–87) or T. W. Anderson (1971, pp. 338–344).

If there is any possibility of correlation existing for greater lags, one would also want to examine the pairs (y_i, y_{i+j}), $i = 1, \cdots, n-j$, for $j > 1$ and/or compute r_2, r_3, etc.

1.3.3. Correction

The best hope is for n to be large enough to permit substituting r_1 for ρ_1 in the variance and correcting the denominator of the t statistic. That is, for large n,

$$\frac{\sqrt{n}(\bar{y} - \mu)}{s\sqrt{1 + 2r_1}} \approx N(0, 1). \tag{1.52}$$

With considerable loss in efficiency, one can divide the data into g consecutive groups with k consecutive observations in each group $(n = g \cdot k)$ and then use the group averages as g approximately independent data points. By grouping, the serial correlation has been reduced to ρ_1/k approximately, but the number of observations has also been reduced by the factor $\frac{1}{k}$.

The sign test and signed-rank test cannot rescue us in this case. In fact, they are in almost as much trouble as the t test. An excellent paper on this topic is by Gastwirth and Rubin (1971).

Letting $\mu = 0$ for notational simplicity the asymptotic variance of the sign statistic is

$$\text{AVar}(S) = n\left(\frac{1}{4} + 2\,\text{Cov}(I\{y_i > 0\},\ I\{y_{i+1} > 0\})\right), \tag{1.53}$$

and, similarly, the asymptotic variance of the signed-rank statistic is

$$\begin{aligned}\text{AVar}(SR_+) = n^3\bigg(\frac{1}{12} + 2\,\text{Cov}(I\{y_i + y_j > 0\},& \\ I\{y_{i+1} + y_k > 0\})\bigg),&\end{aligned} \tag{1.54}$$

where j and k are taken to be far enough removed from i, $i+1$ and each other so as to index uncorrelated observations. Transformation of the positive quadrant to a wedge-shaped region for independent coordinates easily gives

$$\mathrm{Cov}(I\{y_i > 0\},\ I\{y_{i+1} > 0\}) = \frac{\sin^{-1}\ \rho_1}{2\pi}, \tag{1.55}$$

and, since $\mathrm{Cov}(y_i + y_j, y_{i+1} + y_k) = \rho_1/2$,

$$\mathrm{Cov}(I\{y_i + y_j > 0\},\ I\{y_{i+1} + y_k > 0\}) = \frac{\sin^{-1}(\rho_1/2)}{2\pi}. \tag{1.56}$$

These combine with (1.53) and (1.54) to make

$$\begin{aligned} \mathrm{AVar}(S) &= n\left(\frac{1}{4} + \frac{\sin^{-1}\ \rho_1}{\pi}\right), \\ \mathrm{AVar}(SR_+) &= n^3\left(\frac{1}{12} + \frac{\sin^{-1}(\rho_1/2)}{\pi}\right). \end{aligned} \tag{1.57}$$

Since

$$\frac{4}{\pi}\sin^{-1}\ \rho_1 \le \frac{12}{\pi}\sin^{-1}\left(\frac{\rho_1}{2}\right) \le 2\rho_1 \tag{1.58}$$

for $\rho_1 \ge 0$, the effect of positive ρ_1 is the greatest on t and the least on S, but still the effect on S can be appreciable. For instance, with $\rho_1 = \frac{1}{3}$ the limiting standard deviation of the sign test is .6 instead of .5, so for a reported P value of .025 the actual P value would be .051, double the reported value. For the signed-rank test the actual value would be .063.

Gastwirth and Rubin study more general forms of serial correlation for Gaussian processes and for processes with two-tailed exponential distributions. In all cases studied the sign and signed-rank statistics are not appreciably better than the t statistic.

Exercises.

1. Show that the normal theory likelihood ratio test of $H_0 : \mu = \mu_0$ vs. $H_1 : \mu \neq \mu_0$ is equivalent to the two-sided t test.

2. Use the result in (1.11) to show that the asymptotic correlation between $\bar{y}$ and s^2 for $y_1, \cdots, y_n$ independently, identically distributed is

$$\frac{\gamma_1}{(\gamma_2 + 2)^{1/2}},$$

where γ_1 and γ_2 are the population skewness and kurtosis.

3. Show that the Tukey representation (1.32) and (1.33) for SR_+ is correct.

4. Show that for independently, identically, continuously distributed $y_1, \cdots, y_n$

$$\text{Var}(SR_+) = \frac{n(n+1)(2n+1)}{24}.$$

5. For $y_1, \cdots, y_n$ identically distributed with $\text{Var}(y_i) = \sigma^2$, $\text{Cov}(y_i, y_{i+1}) = \rho_1 \sigma^2$, and $\text{Cov}(y_i, y_{i+j}) = 0$, $j \neq 0, 1$, show that for the sample mean $\bar{y}$ and variance s^2

   (a) $\text{Var}(\bar{y}) = \frac{\sigma^2}{n}\left[1 + 2\rho_1\left(1 - \frac{1}{n}\right)\right]$,

   (b) $E(s^2) = \sigma^2\left(1 - \frac{2\rho_1}{n}\right)$.

6. In an experiment at Stanford Medical Center, donor blood was collected into bags containing ACD (an anticoagulant acid citrate dextrose solution) and others containing ACD plus adenine to investigate whether the addition of adenine would better preserve the cryoprecipitates.* The amounts of AHG (antihemophilic gobulin) in donor paired bags were determined at the

---

* Summary Report RFP NHI-67-14, "Effect of ACD-adenine anticoagulant on *in vitro* and *in vivo* potency of cryoprecipitates" by J. G. Pool, Division of Hematology, Stanford University, for the National Heart Institute.

time of administration to 12 hemophilic patients.

ACD:	58.5	82.6	50.8	16.7	49.5	26.0
ACD+A:	63.0	48.4	58.2	29.3	47.0	27.7
ACD:	56.3	35.7	37.9	53.3	38.2	37.1
ACD+A:	22.3	43.0	53.3	49.5	41.1	32.9

Run a t test for the hypothesis of no adenine effect.

7. For the data in Exercise 6 construct a probit plot of differences. Do you think the normality assumption is satisfied?

8. Consider the differences in Exercise 6.

   (a) Compute the median and run a sign test.

   (b) Compute the Hodges-Lehmann estimator and run a signed-rank test.

   (c) Compute a trimmed mean and run a t test with winsorized standard deviation by trimming two data points from each tail.

   Which of these estimators and associated tests, or the mean and t test of Exercise 6, is most appropriate to report for these data?

9. Consider the 16 differences (i.e., -12.7, 18.6, etc.) in the paired data of Exercise 11 for Chapter 3 to be independent. Test the hypothesis of no difference in the tritiated thymidine levels between air and 0_2-exposed mice. Select the test you consider most appropriate, and give the reason(s) for your selection.

Chapter 2

TWO SAMPLES

The previous chapter dealt with the comparison of a sample and a theoretical parameter. When the theoretical parameter is a control or standard value, this value is often not known precisely under the particular conditions of the experiment, so the investigator also obtains a series of control observations. If the experimental and control observations are paired on nuisance characteristics in order to eliminate their effects, then individual differences should be computed for each pair, and the problem remains a one sample problem of comparing the mean difference with zero. When it is not necessary to pair the experimental and control series, the problem becomes a two sample problem.

Other problems in which both sets of data would be called experimental arise as well. The criterion for handling them as one or two sample problems is whether there is any natural pairing between the data sets which should be taken into account in the analysis.

2.1. Normal Theory.

Let $y_{11}, \cdots, y_{1n_1}$ be independently distributed as $N(\mu_1, \sigma_1^2)$, and let $y_{21}, \cdots, y_{2n_2}$ be independently distributed as $N(\mu_2, \sigma_2^2)$. The two samples are assumed to be independent of each other as well. The null hypothesis is customarily $H_0 : \mu_1 = \mu_2$, and the alternative is $H_1 : \mu_1 \neq \mu_2$ or $H_1 : \mu_1 > \mu_2$.

In order to mathematically derive a test the severe assumption

$\sigma_1^2 = \sigma_2^2 = \sigma^2$ is imposed on the model. Under this condition of equal variances the likelihood ratio test of the two-sided alternative leads to the t statistic

$$t = \frac{\bar{y}_1 - \bar{y}_2}{s\sqrt{\frac{1}{n_1} + \frac{1}{n_2}}}, \tag{2.1}$$

where $\bar{y}_i = \sum_{j=1}^{n_i} y_{ij}/n_i$, $i = 1, 2$, and s^2 is the pooled variance

$$s^2 = \frac{1}{n_1 + n_2 - 2}\left[\sum_{j=1}^{n_1}(y_{1j} - \bar{y}_1)^2 + \sum_{j=1}^{n_2}(y_{2j} - \bar{y}_2)^2\right]. \tag{2.2}$$

Under H_0, (2.1) has a t distribution with $n_1 + n_2 - 2$ df so a one-tailed P value is given by $P\{t_{n_1+n_2-2} > t\}$. The two-sided P value would add the areas in both tails.

For confidence intervals the pivotal statistic is

$$t = \frac{(\bar{y}_1 - \bar{y}_2) - (\mu_1 - \mu_2)}{s\sqrt{\frac{1}{n_1} + \frac{1}{n_2}}}, \tag{2.3}$$

so a two-sided $100(1 - \alpha)\%$ confidence interval for $\mu_1 - \mu_2$ is

$$\mu_1 - \mu_2 \in \bar{y}_1 - \bar{y}_2 \pm t_{n_1+n_2-2}^{\alpha/2} s\sqrt{\frac{1}{n_1} + \frac{1}{n_2}}, \tag{2.4}$$

where $t_{n_1+n_2-2}^{\alpha/2}$ is the upper $100(\alpha/2)$ percentile of the t distribution with $n_1 + n_2 - 2$ df. Though infrequently used, a one-sided interval could also be constructed.

2.2. Nonnormality.

2.2.1. Effect

The effects of nonnormality on (2.1) are similar but not identical to the effects on the one sample t statistic. The reader should therefore be familiar with Section 1.2.1 before pursuing the discussion here.

As in the one sample case the t analysis is validated in the limit by the central limit theorem. For small samples, however, the skewness and to a lesser extent the kurtosis of the populations can have some effect. Continue to assume $\sigma_1^2 = \sigma_2^2$ since the effect of unequal variances is examined in Section 2.3.1, but let $\gamma_1(y_1)$, $\gamma_1(y_2)$ and $\gamma_2(y_1)$, $\gamma_2(y_2)$ be the skewness and kurtosis parameters of the y_1 and y_2 populations. Then the story of nonnormality is pretty much contained in the leading terms of the expansions for the first three moments of t, which were derived by Geary (1947) and Gayen (1950b):

$$E(t) \cong \frac{1}{\nu_1^{1/2}}\left[-\frac{1}{2}(\gamma_1(y_1) - \gamma_1(y_2))\frac{1}{\nu_2}\right],$$

$$\begin{aligned}\mathrm{Var}(t) \cong \frac{1}{\nu_1}\Bigg[\left(1 + \frac{2}{\nu_2}\right)\nu_1 &+ \frac{7}{4}(\gamma_1(y_1) - \gamma_1(y_2))^2\frac{1}{\nu_2^2} \\ &+ (\gamma_2(y_1) - \gamma_2(y_2))(n_1 - n_2)\frac{\nu_1}{\nu_2^2}\Bigg],\end{aligned} \tag{2.5}$$

$$E(t - E(t))^3 \cong \frac{1}{\nu_1^{3/2}}\left[\frac{\gamma_1(y_1)}{n_1^2} - \frac{\gamma_1(y_2)}{n_2^2} - 3(\gamma_1(y_1) - \gamma_1(y_2))\frac{\nu_1}{\nu_2}\right],$$

where $\nu_1 = (1/n_1) + (1/n_2)$, $\nu_2 = n_1 + n_2 - 2$.

In many experimental applications the assumption that $\gamma_1(y_1) \cong \gamma_1(y_2)$ and $\gamma_2(y_1) \cong \gamma_2(y_2)$ would seem warranted. If this is the case, then the expressions in (2.5) clearly show that the kurtosis parameters have little effect on the t statistic and when the sample sizes are approximately equal (i.e., $n_1 \cong n_2$) the skewness parameters cancel each other approximately. Thus for equal sample sizes the t statistic is more robust in the two sample problem than in the one sample problem. It therefore behooves the investigator to perform a balanced experiment if at all possible.

These theoretical considerations are supported by the Monte Carlo work of Pearson (1929) for $\gamma_1(y_1) = \gamma_1(y_2)$ between 0 and .7, $\gamma_2(y_1) = \gamma_2(y_2)$ between $-.5$ and 4, and samples sizes in the range 5 to 20, and of Pearson and Please (1975) for $\gamma_1(y_1) = \gamma_1(y_2)$ between 0 and .8, $\gamma_2(y_1) = \gamma_2(y_2)$ between -1 and 1.4, and equal sample sizes between 10 and 25.

For n_1 and n_2 not approximately equal, the skewness of the mean with the smaller sample size dominates the numerator of the t statistic. Since s^2 is a weighted average of the two sample variances, i.e.,

$$s^2 = \frac{(n_1 - 1)s_1^2 + (n_2 - 1)s_2^2}{n_1 + n_2 - 2}, \tag{2.6}$$

where

$$s_i^2 = \frac{1}{n_i - 1}\sum_{j=1}^{n_i}(y_{ij} - \bar{y}_i)^2, \quad i = 1, 2, \tag{2.7}$$

the variance for the larger sample tends to dominate the denominator of the t statistic. Since the dominating mean and dominating variance are independent, there is less dependence between numerator and denominator in the two sample case than in the one sample, and the skewness of t remains in the direction of the skewness of the mean with smaller sample size. Recall that in the one sample problem the direction of skewness was reversed by the correlation between numerator and denominator. Even for n_1 and n_2 not approximately equal the kurtosis has only a minor effect on t.

More serious distortion of the P values can occur when $\gamma_1(y_1)$ does not approximately equal $\gamma_1(y_2)$. The leading terms do not cancel out in this case even for equal sample sizes. Fortunately, this case does not seem to occur frequently. When it does occur, it is questionable whether an analysis of the mean values is an appropriate comparison for the two populations with quite different shapes.

Although the P value from a t statistic is reasonably trustworthy, it still may not be the best statistic to use for nonnormal distributions. Sharper results in terms of increased power or smaller P values may be obtainable through alternative parametric or nonparametric procedures.

Just as in the one sample case, outliers can distort the mean difference and the t statistic. Their major impact on the statistic (2.1) is to inflate the variance estimate (2.2) and thereby depress the value and corresponding statistical significance of (2.1).

2.2.2. Detection

For a full discussion of detecting nonnormality the reader is referred to Section 1.2.2. One sample methods can be applied to each of the two samples. Probit plots of each sample are a worthwhile way to scrutinize the data.

The presence of more than one sample does not substantially alter the problem except through the advent of variance stabilizing transformations. Their use is described in Section 2.3. The connection between variance stabilizing transformations and nonnormality is mainly empirical. It often happens in practice that the transformation that best stabilizes the variance also improves the appearance of normality in the data. Skewed long tails in the samples affect both the variances and the probit plots. Thus methods for detecting and correcting inequality of variance are in a broad sense also methods for detecting and correcting nonnormality.

As in the one sample problem, outliers can be detected as well through probit plots.

2.2.3. Correction

Transformations As mentioned previously, transformations can be very useful in improving the normality of the data. For positive data the *logarithmic* and *square root transformations* are the most frequently employed because of easy access to tables, special keys on electronic calculators, and readily available commands on large computers. When some of the data take values close to zero, addition of a small constant to each observation before it is transformed may increase the effectiveness of the transformation. Other transformations are of course possible, and for a full discussion the reader is referred to Section 1.2.3.

Selection of a transformation is still mainly guesswork and experience, or is suggested by examination of the variances (see Sections 2.3.2 and 2.3.3). Probit plots of the transformed data are a worthwhile check on the wisdom of the selection.

Transformations are not customarily useful in correcting for outliers. Nonparametric techniques and robust estimators are better suited for handling outliers.

Nonparametric Techniques As in the one sample problem there are three principal nonparametric tests. The *two sample median test* is the two sample analog of the sign test. For reasons not entirely clear it is not used with the frequency of the sign test. The two sample Wilcoxon test is by far the more common. Nevertheless, the two sample median test is a quick, easy, and robust test. To execute the test combine the two samples into one and calculate the median m_c of the combined sample. For $n_1 + n_2$ odd, the median is an observation from one of the samples; for $n_1 + n_2$ even, it is the average of the middle observations. Separate the data into the original samples and within each sample count the number of observations above and below m_c. The counts can be neatly summarized in a 2×2 table:

	$<$	$m_c <$	
Sample 1	a	b	a+b
Sample 2	c	d	c+d
	a+c	b+d	N=a+b+c+d

(2.8)

Observations with values equal to m_c are analogous to ties with zero in the sign test and are a source of annoyance. It is hoped that there are few of them. My preference is to exclude the values tied with m_c including the value m_c itself when $n_1 + n_2$ is odd. A conservative approach would place all the ties in each sample in the direction opposite to significance; i.e., make $ad - bc$ as close to zero as possible. For large number of ties the reader is left to decide for himself or herself. There does not seem to have been as extensive a study of ties in this two sample problem as in the one sample case.

Once the 2×2 table has been created, the analysis can proceed as for a 2×2 contingency table. The quickest analysis is to compute the χ^2 *statistic*

$$\frac{N\left(|ad - bc| - \frac{N}{2}\right)^2}{(a+b)(c+d)(a+c)(b+d)}. \tag{2.9}$$

Under the null hypothesis of no difference between the populations this has a limiting χ^2 distribution with one df as $n_1,\ n_2 \to \infty$. Various rules of thumb exist for how large n_1 and n_2 have to be for the χ^2 approximation to be valid. For $\min\{n_1, n_2\} \geq 10$ and $\min\{a, b, c, d\} \geq 2$, I feel the χ^2 approximation is quite good for practical purposes. The sometimes suggested rule that the expected number in each cell should be at least 5 is unnecessarily conservative.

The P value computed from the upper tail of the χ^2 distribution with one df is a two-sided P value since the test rejects when the first sample has larger values than the second and vice versa. For a one-sided P value take the square root of (2.9) and assign it a $+$ or $-$ sign depending on whether population 1 or 2 has larger values. Tables of the normal distribution can then be used to obtain a one-sided P value.

There is disagreement over whether it is best to include the *Yates'* (1934) *continuity correction* $N/2$ in the numerator of (2.9). Since the aim here is to accurately approximate the P value for the exact analysis to be discussed next, its use is justified. When many values of the χ^2 statistic are being computed as a screening device for detecting possible differences in large sets of data, then it is best to leave it out. The uncorrected statistic has a false positive rate closer to the nominally stated α. Also when pooling separate χ^2 with one df as in cooperative studies, it may be best to use the uncorrected χ^2. For a full discussion of the controversy the reader is referred to Mantel and Greenhouse (1968), Grizzle (1967, 1969), and Conover (1974) with appended comments.

For a finer analysis with small sample sizes there is *Fisher's* (1934) *exact test*.* Under the null hypothesis the conditional distribution of the table entries given the four marginal totals is hypergeometric; i.e.,

$$\begin{aligned} P\{a,b,c,d \mid a+b, c+d, a+c, b+d\} &= \frac{\binom{a+b}{a}\binom{c+d}{c}}{\binom{N}{a+c}}, \\ &= \frac{(a+b)!(c+d)!(a+c)!(b+d)!}{N!a!b!c!d!}. \end{aligned} \tag{2.10}$$

A one-tailed P value is obtained by summing the probabilities (2.10) for each table equal to and more extreme than the observed with the same marginal totals. For example, if the observed table is

5	2
3	7

, (2.11)

then one would sum the probabilities for the tables

5	2
3	7

6	1
2	8

7	0
1	9

. (2.12)

It is not always clear how to obtain a two-sided P value. The

---

* This test was also proposed by Irwin (1935); in addition see Yates (1934).

remainder of the sequence of tables is

0	7
8	2

1	6
7	3

2	5
6	4

(2.13)

3	4
5	5

4	3
4	6

.

In this case the two tables on the left would be considered more extreme than the observed. In other examples, however, some criterion may have to be introduced to measure the degree of disagreement with the null hypothesis of questionable tables. One criterion might be the size of the χ^2 statistic (2.9), or equivalently, the size of $|ad-bc|$. Another would be the size of the probability (2.10). It would be unfortunate if the scientific conclusion rested on which criterion were selected. When $a + b = c + d$, there is no ambiguity because of the symmetry in the sequence. A convention, which is sometimes used and avoids the aforementioned dilemma when $n_1 \neq n_2$, is to simply double the one-tailed P value to get a two-tailed P value.

Computation of the probability (2.10) is usually easy. The numbers are usually not large (otherwise the χ^2 approximation could be used) and a great deal of cancellation occurs. Some of the better electronic calculators have special keys for $N!$, and some programmable ones have programs for calculating (2.10). Once one probability has been computed, the values for neighboring tables can be generated quickly by multiplication and division with the appropriate integers to give the new factorials.

Finney et al. (1963) give a set of tables of critical values for Fisher's exact test. The tables are easy to use but unfortunately they are not always readily available.

The most popular two sample test next to the t test is the

Wilcoxon (1945) *rank test*. Its asymptotic efficiency compared to the t is quite high for the normal distribution (i.e., $3/\pi \cong .95$), and it is more efficient than the t for many heavy-tailed distributions. Compared to the t its asymptotic efficiency never drops below .864. Outliers have no appreciable effect on it. It is quick and easy to compute, and good tables are readily available.

The Wilcoxon statistic can be computed in either of two ways.

One method depends on ranking. Combine the two samples into one set of $n_1 + n_2$ observations. Order the observations from smallest to largest $y_{(1)} \leq y_{(2)} \leq \cdots \leq y_{(n_1+n_2)}$, and assign i to the ith largest observation. Let R_1 be the sum of the ranks attached to the observations from the first sample and, similarly, let R_2 be the rank sum for the second sample. The Wilcoxon statistic is either R_1 or R_2, or possibly $R_1 - R_2$ when $n_1 = n_2$. Since

$$R_1 + R_2 = \frac{(n_1 + n_2)(n_1 + n_2 + 1)}{2}, \tag{2.14}$$

any one of these statistics contains all the information on the rank sums.

The Mann-Whitney (1947) form of the Wilcoxon statistic is

$$U = \sum_{i=1}^{n_1} \sum_{j=1}^{n_2} I\{y_{1i} > y_{2j}\}, \tag{2.15}$$

where

$$I\{y_{1i} > y_{2j}\} = \begin{cases} 1 & \text{if } y_{1i} > y_{2j}, \\ 0 & \text{if } y_{1i} < y_{2j}. \end{cases} \tag{2.16}$$

This can usually be quickly computed by taking each y_{1i} observation and scanning the second sample to count how many y_{2j} values are smaller than y_{1i}.

The counting method (2.15) is related to the ranking procedure through

$$R_1 = \frac{n_1(n_1+1)}{2} + U. \tag{2.17}$$

The argument for (2.17) is simple. If all the y_{1i} preceded all the y_{2j}, the rank sum R_1 would be $n_1(n_1+1)/2$ and U would be zero. Each time a y_{2j} comes before a y_{1i} it increases the rank of y_{1i} by one and the sum U by one.

The easiest way of handling ties is to assign an average rank to each of the tied observations. For example, if $y_{11} = .3$, $y_{12} = .6$, $y_{21} = 1.1$, $y_{22} = .6$, then y_{12} and y_{22} would each receive the average rank score 2.5. This is equivalent to expanding the definition of the indicator function in (2.16) to include

$$I\{y_{1i} > y_{2j}\} = \frac{1}{2} \quad \text{if } y_{1i} = y_{2j}. \tag{2.18}$$

For small numbers of ties the ordinary tables and large sample approximations can be used without alteration with no serious effect on the inference. For a moderate number of ties the tables can still be used to get an idea of the P value, but one must be aware that the variability of the Wilcoxon statistic has been reduced. A correction to the variance of the Wilcoxon statistic, conditional on the pattern of ties, can be made [see (2.20)] but the ties must be substantial before the correction reaches appreciable magnitude. Numerous ties can, of course, leave the inference in doubt. An excellent paper on the effect of ties on the Wilcoxon statistic is Klotz (1966).

Good tables of the Wilcoxon statistic are usually readily available. Many textbooks contain abbreviated tables in their appendices. Owen (1962) and Pearson and Hartley (1972) each contain a set. When using whatever tables are available, one must check precisely what is being tabled. Some give tail probabilities or critical values for R_1, others for U.

Asymptotically, U (and R_1 or R_2) has a normal distribution.

Under the null hypothesis of no difference between the populations its exact mean and variance are

$$E(U) = \frac{n_1 n_2}{2},$$
$$\mathrm{Var}(U) = \frac{n_1 n_2 (N+1)}{12}, \tag{2.19}$$

where $N = n_1 + n_2$. The large sample approximation is very good by the time n_1 and n_2 are at least 10, and it can be used with impunity for somewhat smaller samples provided neither one is quite small.

When ties are present and are handled by means of (2.18), the exact mean and variance of U, conditional on the pattern of ties, can be calculated. The conditional mean of U is still $n_1 n_2/2$. For the calculation of the variance, let $z_1, \cdots, z_m$ be the distinct values in the combined sample of y_{1i} and y_{2j}, and let $t_1, \cdots, t_m$ be the numbers of observations that equal each of these values. In the case of an observation with no other equal to it, $t_i = 1$. Then the conditioanl variance of U is

$$\begin{aligned} \mathrm{Var}(U \mid t_1, \cdots, t_m) &= \frac{n_1 n_2}{12}\left[N + 1 - \frac{\sum_{i=1}^m (t_i^3 - t_i)}{N(N-1)}\right], \\ &= \frac{n_1 n_2 (N+1)}{12}\left[1 - \frac{\sum_{i=1}^m (t_i^3 - t_i)}{N^3 - N}\right]. \end{aligned} \tag{2.20}$$

Experience will teach the reader that the tie correction factor $1 - [\sum_{i=1}^m (t_i^3 - t_i)/(N^3 - N)]$ does not become substantially less than one very fast.

From (2.17) the variance of R_1 is the same as that of U. The mean of R_1 differs from that of U by the additive factor $n_1(n_1+1)/2$.

Before leaving the Wilcoxon statistic, several remarks are in order.

First, unlike the one sample Wilcoxon signed-rank test, there is no assumption of symmetry of the underlying distributions. Symmetry does not play a role in the two sample problem.

Second, the statistic U/n_1n_2 is estimating the probability $P\{y_1 > y_2\}$ in the continuous case, and $P\{y_1 > y_2\} + \frac{1}{2}\, P\{y_1 = y_2\}$ when the distributions have discrete mass points. For the continuous case the test will be consistent against any alternative for which $P\{y_1 > y_2\}$ differs from $\frac{1}{2}$. The statistic U/n_1n_2 is a special case of a two sample U-statistic in the sense of Hoeffding (1948).

Third, the estimator for the difference in location of the two populations associated with the Wilcoxon rank statistic is the two sample *Hodges-Lehmann* (1963) *estimator*. This estimate $\hat{\Delta}_{HL}$ is the median of the collection of n_1n_2 values $\{y_{1i} - y_{2j},\ i = 1, \cdots, n_1,\ j = 1, \cdots, n_2\}$.

A confidence interval for the true difference Δ in location of the two populations can be constucted from the Wilcoxon statistic.* In the Mann-Whitney form the confidence interval consists of all values of Δ for which $U(\Delta) = \sum_{i=1}^{n_1} \sum_{j=1}^{n_2} I\{y_{1i} - \Delta > y_{2j}\}$ does not differ significantly from the null mean $n_1n_2/2$. This is tedious to construct numerically, but a graphical method due to Moses (see Walker and Lev, 1953, Chapter 18) greatly simplifies the calculation. Plot the n_1n_2 points (y_{1i}, y_{2j}), $i = 1, \cdots, n_1$, $j = 1, \cdots, n_2$ on a sheet of graph paper. Let $u^{\alpha/2}$ be the lower tail critical point for the U statistic based on n_1, n_2 observations; i.e., $u^{\alpha/2}$ is the largest integer such that $P[U \leq u^{\alpha/2} \mid H_0\} \leq \alpha/2$. In large samples

$$u^{\alpha/2} \cong \frac{n_1n_2}{2} - \frac{1}{2} - z^{\alpha/2}\left[\frac{n_1n_2(n_1 + n_2 + 1)}{12}\right]^{1/2}, \qquad (2.21)$$

where $1/2$ is a continuity correction and $z^{\alpha/2}$ is the upper $100(\alpha/2)$ percentile of a normal distribution. Slide a 45° line along the y_1 axis until $u^{\alpha/2}$ points lies to the right of the line and one lies on it; call

---

* The underlying assumption is that the shapes of the two distribitions are the same except for their location. Thus Δ is the difference between the means or the differences between the medians.

the y_1 value where the line crosses the y_1 axis Δ_U. Similarly, let Δ_L be the y_1 value at which a 45° line through $(y_1, 0)$ has $u^{\alpha/2}$ points to the left of it and one on it. The interval between Δ_L and Δ_U is the confidence interval for Δ. The procedure is illustrated in Figure 2.1 with $n_1 = n_2 = 3$, $u^{\alpha/2} = 1$.

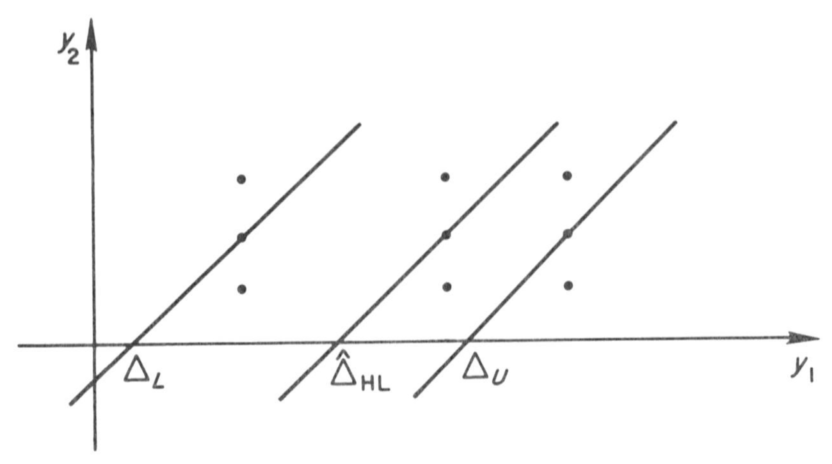

Figure 2.1

The third and final nonparametric test to be mentioned is *Pitman's* (1937) *permutation test*. It illustrates the general principle of permutation inference. Select a statistic that should be sensitive to the type of alternative hypothesis of interest. For the two sample problem, $\bar{y}_1 - \bar{y}_2$ is a prime candidate. Compute the value of $\bar{y}_1 - \bar{y}_2$ for the observed samples, and also the $\binom{n_1+n_2}{n_1}$ hypothetical values obtainable by dividing the combined sample of size $n_1 + n_2$ into all possible pairs of subsets of sizes n_1 and n_2. Under the null hypothesis of no difference between the populations the conditional probability, given the combined sample, of each possible pair of samples is $\binom{n_1+n_2}{n_1}^{-1}$. If the observed $\bar{y}_1 - \bar{y}_2$ lies far out in the tail(s) of the range of possible values, then it is judged significant. The positive one-tailed P value is the number of values $\bar{y}_1 - \bar{y}_2$ greater than or equal to the observed divided by $\binom{n_1+n_2}{n_1}$. A two-tailed P value uses $|\bar{y}_1 - \bar{y}_2|$.

The permutation test is not used much because it is computationally unwieldy except for small sample sizes or for obviously very extreme values of $\bar{y}_1 - \bar{y}_2$. Large electronic computers aid in this problem, but even they can be taxed if n_1 and n_2 are moderately large. Generation of random permutations in the computer to estimate the P value is a solution to this dilemma, but for the simple two sample problem it seems simpler to use something else, like the Wilcoxon or t statistics. Asymptotically, the permutation test is equivalent to the t test.

Instead of random permutations the *bootstrap method* of sampling could be used; see Efron (1979, 1982).

Robust Estimation Discussion of robust estimation in the two sample problem is limited here to just trimmed means. For more details on robust estimators in general the reader is referred to Section 1.2.3, "Robust Estimation."

As in the one sample problem, there is an underlying assumption that the cdf for each population is symmetric about its median. Without this assumption the rationale for the estimators and the distribution theory break down. If the assumption appears to be grossly violated, the statistician may be able to first transform the data to achieve better symmetry.

In the two sample problem one simply repeats twice what is done in the one sample problem and pools the variances. Specifically, let δ be the trimming fraction, where it is assumed that δn_1 and δn_2 are integers. For $i = 1, 2$, let

$$\bar{y}_{Ti} = \frac{1}{(1-2\delta)n_i} \sum_{j=\delta n_i+1}^{n_i-\delta n_i} y_{i(j)}, \tag{2.22}$$

where $y_{i(1)} \leq y_{i(2)} \leq \cdots \leq y_{i(n_i)}$ are the order statistics for the ith sample, and let

$$s_{Wi}^2 = \frac{1}{(1-2\delta)^2(n_i - 1)}\Bigg[\delta n_i(y_{i(\delta n_i+1)} - \bar{y}_{Wi})^2 + \sum_{j=\delta n_i+1}^{n_i-\delta n_i} (y_{i(j)} - \bar{y}_{Wi})^2 + \delta n_i(y_{i(n_i-\delta n_i)} - \bar{y}_{Wi})^2\Bigg], \tag{2.23}$$

where

$$\bar{y}_{Wi} = \frac{1}{n_i}\left[\delta n_i\, y_{i(\delta n_i+1)} + \sum_{j=\delta n_i+1}^{n_i-\delta n_i} y_{i(j)} + \delta n_i\, y_{i(n_i-\delta n_i)}\right]. \tag{2.24}$$

Then the pooled sample variance is

$$s_W^2 = \frac{(n_1-1)s_{W1}^2 + (n_2-1)s_{W2}^2}{n_1+n_2-2}, \tag{2.25}$$

and the appropriate trimmed t statistic for testing $H_0 : F_1 = F_2$ is

$$t_T = \frac{\bar{y}_{T1} - \bar{y}_{T2}}{s_W\sqrt{\frac{1}{n_1} + \frac{1}{n_2}}}. \tag{2.26}$$

Yuen and Dixon (1973) have provided evidence that (2.26) is approximately distributed as a t distribution with $(1-2\delta)(n_1+n_2)-2$ df.

The pooled variance (2.25) and the t statistic (2.26) are based on the assumption that the two population cdfs $F_1$and F_2 are identical (and symmetric) except for a location shift. Without the identity assumption the problem is analogous to the case of $\sigma_1^2 \neq \sigma_2^2$ (see Section 2.3). It should be mentioned that for this problem there is a statistic utilizing trimmed means with unpooled variances that is analogous to Welch's approximate t' statistic (see Section 2.3.3, "Other Tests"). For details the reader is referred to Yuen (1974).

2.3. Unequal Variances.

The model is that the y_{ij}, $i = 1, 2$, $j = 1, \cdots, n_i$, are independently distributed as $N(\mu_i,\ \sigma_i^2)$ without the assumption $\sigma_1^2 = \sigma_2^2$.

2.3.1. Effect

Under this model

$$\bar{y}_1 - \bar{y}_2 \sim N\left(\mu_1 - \mu_2,\ \frac{\sigma_1^2}{n_1} + \frac{\sigma_2^2}{n_2}\right), \tag{2.27}$$

and this is also true asymptotically without the assumption of normality. * Let $n_1/n_2 \to R$ as $n_1, n_2 \to \infty$. Then

$$s^2 = \frac{n_1 - 1}{n_1 + n_2 - 2} s_1^2 + \frac{n_2 - 1}{n_1 + n_2 - 2} s_2^2 \xrightarrow{p} \frac{R}{1+R}\sigma_1^2 + \frac{1}{1+R}\sigma_2^2, \tag{2.28}$$

and this too is true asymptotically for nonnormal distributions. Thus with or without the assumption of normality,

$$\begin{aligned} t &= \frac{(\bar{y}_1 - \bar{y}_2) - (\mu_1 - \mu_2)}{s\sqrt{\frac{1}{n_1} + \frac{1}{n_2}}}, \\ &= \frac{(\bar{y}_1 - \bar{y}_2) - (\mu_1 - \mu_2)}{\sqrt{\frac{\sigma_1^2}{n_1} + \frac{\sigma_2^2}{n_2}}} \times \frac{\sqrt{\frac{\sigma_1^2}{n_1} + \frac{\sigma_2^2}{n_2}}}{s\sqrt{\frac{1}{n_1} + \frac{1}{n_2}}} \\ &\xrightarrow{d} N\left(0,\ \frac{1}{\frac{R}{1+R}\sigma_1^2 + \frac{1}{1+R}\sigma_2^2} \times \frac{\sigma_1^2 + R\,\sigma_2^2}{1+R}\right). \end{aligned} \tag{2.29}$$

The asymptotic variance of t, instead of being equal to one, is

$$\mathrm{AVar}(t) = \frac{\theta + R}{R\theta + 1}, \tag{2.30}$$

where $\theta = \sigma_1^2/\sigma_2^2$.

How do different values of θ effect AVar(t), and how, in turn, does this affect the large sample inference?

---

* "$\sim$" denotes "is distributed as."

Note first that when $R = 1$ (i.e., $n_1 = n_2$), AVar(t) = 1. This means that when the sample sizes are equal, inequality of variance does not affect the inference asymptotically. If the sample sizes are nearly equal, the t test can tolerate large disparities in the variances (viz., ratios of 4 and up) without showing major ill effects. Thus it pays to balance the experiment as closely as possible.

Consider another case: $\theta = 2$, $R = 2$. Here the variance for the first population is twice as large as for the second, but the first population also has twice as large a sample. In this case AVar(t) = .8 so the asymptotic standard deviation is approximately .9 instead of 1. The effect on the P value is not large. A reported two-sided P value of .05 would in actuality be $P = .03$. In his Table 10.2.3 Scheffé (1959, p. 340) gives more examples to illustrate the effects on P for varying θ and R.

The worst situation is where the variance of population 1 is very much larger than for population 2 (i.e., $\sigma_1^2 >> \sigma_2^2$) and the sample size for the first population is much smaller (i.e., $n_1 << n_2$). The least information is available on the larger variance. In this case t would be handled as though it had $n_1 + n_2 - 2$ df, which would be large because of n_2, whereas t is approximately behaving like

$$\frac{\bar{y}_1 - \mu_1}{\left[\left(\frac{n_1}{n_1+n_2}\right)s_1^2 + \sigma_2^2\right]^{1/2} \left[\frac{1}{n_1}\right]^{1/2}}, \tag{2.31}$$

because $\bar{y}_2 - \mu_2 \cong 0, s_2^2 \cong \sigma_2^2$, $n_2/(n_1 + n_2) \cong 1$, and $1/n_2 \cong 0$. If $n_1 s_1^2/(n_1 + n_2)$ is small relative to σ_2^2, the ratio in (2.31) behaves like a normal variable with variance σ_1^2/σ_2^2 instead of 1, and if $n_1 s_1^2/(n_1 + n_2)$ is large relative to σ_2^2, it behaves like a t variable on $n_1 - 1$ df multiplied by $\sqrt{(n_1 + n_2)/n_1}$. In either case the variability is greater than that hypothesized by a t distribution on $n_1 + n_2 - 2$ df. As an illustrative example from Table 10.2.3 in Scheffé (1959, p. 340), the actual significance level for large n_1, n_2 is .22 instead of .05 when

$\theta = 5$ and $R = 1/5$.

2.3.2. Detection

It is far harder to decide whether σ_1^2 equals σ_2^2 than it is to correct for their inequality. The problem is that the standard textbook test based on s_1^2/s_2^2 having an F distribution is extremely sensitive to departures from normality and cannot be relied upon. Chapter 7 considers this problem in detail, and alternative robust test procedures are described. All involve extra computation. Since the effects on t are not large unless the variance disparity is sizeable and the experiment is badly unbalanced, preliminary tests of $\sigma_1^2 = \sigma_2^2$ seem to be a fruitless pastime. Worrisome differences in the variances that are detectable to the naked eyeball lead one to correct for unequal variances without the intermediate step of deciding whether $\sigma_1^2 = \sigma_2^2$.

2.3.3. Correction

Transformations Transformations are often useful in eliminating inequalities between variances. The analysis is then conducted in the transformed scale, although the results are usually reported in the original scale.

Selection of a transformation can be facilitated by the following simple large sample relationship. Consider a smooth function $g(y)$ of the random variable y. If y is fairly tightly distributed about its mean μ, then in the expansion

$$g(y) = g(\mu) + (y - \mu)\, g'(\mu) + 0((y - \mu)^2) \tag{2.32}$$

the second order term will not be substantial compared with the linear term. Rewriting this as

$$g(y) - g(\mu) \cong (y - \mu)\, g'(\mu) \tag{2.33}$$

suggests that

$$E\{[g(y) - g(\mu)]^2\} \cong \mathrm{Var}(y)[g'(\mu)]^2. \tag{2.34}$$

Since $E[g(y)] \cong g(\mu)$, the term on the left in (2.34) approximates $\mathrm{Var}[g(y)]$ so *

$$SD[g(y)] \cong SD(y)|g'(\mu)|. \tag{2.35}$$

The preceding approximations can be justified asymptotically if $y \xrightarrow{d} N(\mu, \sigma^2)$. This procedure for obtaining $\mathrm{Var}[g(y)]$ in the limit is known as the *delta method.* Two important special cases of (2.35) are the approximations for the logarithmic and square root transformation variances. For $\log y$, $g'(y) = 1/y$; thus

$$SD(\log y) \cong \frac{SD(y)}{\mu}. \tag{2.36}$$

The ratio on the right in (2.36) is the *coefficient of variation* of y. Many measured variables have a constant coefficient of variation, or constant *percent error* as it is sometimes called, in which case the log transform is appropriate. For $\sqrt{y}$, $g'(y) = 1/2\sqrt{y}$; thus

$$SD(\sqrt{y}) \cong \frac{SD(y)}{2\sqrt{\mu}}. \tag{2.37}$$

With Poisson data, the variance equals the mean so the square root transform should stabilize the variances.

The relationship (2.35) is quite helpful when there are two or more samples, for then it is possible to plot s_i vs. $\bar{y}_i$ to see if any empirical relationship holds between the sample standard deviations and means. If, for example, the standard deviation increases as some power of the mean, then (2.35) suggests trying a power transformation with the power increased by one. Fiddling with the transforma-

* "$SD(y)$" denotes the standard deviation of y.

tion by adding a constant to the variables may improve the stability of the variances.

In the two sample problem there are just two points $(s_1, \bar{y}_1)$ and $(s_2, \bar{y}_2)$ so only a little information is available through this procedure. However, if s increases as $\bar{y}$ increases, a power transformation like log or square root may work, whereas if s decreases, a different type such as $g(y) = 1/y$ would be required. Visual inspection of the samples may give some added indication of the proper transformation. Increasing standard deviation with increasing mean is often accompanied in practice by samples skewed to the right with long upper tails. Examination of the upper tails of the samples may shed some light on whether a square root transformation, or the stronger log transformation, is required.

Other Tests The other method of correction is to use a different test. This problem (i.e., two normal populations with $\sigma_1^2 \neq \sigma_2^2$; $H_0 : \mu_1 = \mu_2$) is a classic one in statistical history and is referred to as the Behrens-Fisher problem. Various methods, including fiducial probability, have been proposed for handling it. Scheffé (1970b) nicely summarized the current state of knowledge. In earlier work (1943, 1944) he gave a solution that has an exact t distribution but which is not really suitable for practical work. It employs artificial randomization, and in his 1970 paper Scheffé recommended against its usage. The best solution from the practical point of view is the following approximate one.

The practical procedure is *Welch's t' test.* It uses the statistic

$$t' = \frac{\bar{y}_1 - \bar{y}_2}{\sqrt{\frac{s_1^2}{n_1} + \frac{s_2^2}{n_2}}}. \tag{2.38}$$

Since $s_1^2 \xrightarrow{p} \sigma_1^2$, $s_2^2 \xrightarrow{p} \sigma_2^2$, t' is asymptotically distributed as $N(0, 1)$ when $n_1, n_2 \to \infty$. Thus for large samples the denominator in (2.38) is correctly estimating the standard deviation of the numerator, and this asymptotic convergence is valid even if the populations are non-normal.

The exact distribution of t' under H_0 depends on the unknown σ_1^2 and σ_2^2. Welch (1947, 1949) proposed approximating its distribution by a t distribution with suitably chosen degrees of freedom for small or moderate sample sizes. Welch's approximation for the t' distribution and Satterthwaite's (1946) approximation for the distribution of a linear combination of χ^2 variables employ the same idea. It is to approximate the distribution of the variance combination

$$\frac{s_1^2}{n_1} + \frac{s_2^2}{n_2} \tag{2.39}$$

by the distribution of a χ_ν^2 variable multiplied by σ^2/ν, where σ^2 and ν are chosen so that the first two moments of $\sigma^2\chi_\nu^2/\nu$ agree with the first two moments of (2.39).* In this case

$$E\left(\frac{s_1^2}{n_1} + \frac{s_2^2}{n_2}\right) = \frac{\sigma_1^2}{n_1} + \frac{\sigma_2^2}{n_2}, \tag{2.40}$$

so $\sigma^2 = E(\sigma^2\chi_\nu^2/\nu)$ should be chosen equal to (2.40). The two variances are

$$\operatorname{Var}\left(\frac{s_1^2}{n_1} + \frac{s_2^2}{n_2}\right) = \frac{2\sigma_1^4}{(n_1-1)n_1^2} + \frac{2\sigma_2^4}{(n_2-1)n_2^2}, \tag{2.41}$$

and

$$\operatorname{Var}\left(\frac{\sigma^2}{\nu}\chi_\nu^2\right) = \frac{2\sigma^4}{\nu}. \tag{2.42}$$

Equating (2.42) with (2.41) shows that ν should be chosen to be

$$\frac{\left(\frac{\sigma_1^2}{n_1} + \frac{\sigma_2^2}{n_2}\right)^2}{\frac{1}{n_1-1}\left(\frac{\sigma_1^2}{n_1}\right)^2 + \frac{1}{n_2-1}\left(\frac{\sigma_2^2}{n_2}\right)^2}. \tag{2.43}$$

* "χ_ν^2" denotes a χ^2 variable (or distribution) with ν df.

This still involves the unknown parameters, but s_i^2 can be substituted as an estimate of σ_i^2. This leads to assuming that t' has an approximate t distribution with

$$\hat{\nu} = \frac{\left(\frac{s_1^2}{n_1} + \frac{s_2^2}{n_2}\right)^2}{\frac{1}{n_1-1}\left(\frac{s_1^2}{n_1}\right)^2 + \frac{1}{n_2-1}\left(\frac{s_2^2}{n_2}\right)^2} \tag{2.44}$$

degrees of freedom. P values can then be calculated from t tables with the degrees of freedom equal to the integer nearest $\hat{\nu}$.

It may not be necessary to actually calculate $\hat{\nu}$. A little algebra establishes that

$$\min\{n_1 - 1,\ n_2 - 1\} \le \hat{\nu} \le n_1 + n_2 - 2. \tag{2.45}$$

The extreme df may be sufficient to establish the significance or nonsignificance of the sample. If t' has a high P value even for $n_1 + n_2 - 2$, then the sample difference cannot be statistically significant for $\hat{\nu}$. Similarly, if t' gives a low P value for $\min\{n_1 - 1,\ n_2 - 1\}$, then the difference must be even more significant for $\hat{\nu}$. Results of Hsu (1938) show that, when the populations are normal, use of the min df is a conservative procedure. Namely,

$$P\left\{\left|\frac{(\bar{y}_1 - \mu_1) - (\bar{y}_2 - \mu_2)}{\sqrt{(s_1^2/n_1) + (s_2^2/n_2)}}\right| < t_{\underline{\nu}}^{\alpha/2}\right\} \ge 1 - \alpha, \tag{2.46}$$

where $\underline{\nu} = \min\{n_1 - 1,\ n_2 - 1\}$.

Transformations and Welch's approximate t' test are the procedures I most frequently use to handle unequal variances. The approximate t' test should enjoy the robustness properties of the t test with equal variances, but serious nonnormality of the data may motivate one to use either the trimmed t' test or the nonparametric tests described earlier. The Wilcoxon rank test is also affected by unequal variances, but guesswork and some results of Potthoff (1963) suggest that the effects are small.

2.4. Dependence.

There is little more to be said than what appears in Section 1.3. The effect on $\bar{y}_i$ and s_i of a serial correlation within each sample is, of course, the same as in the one sample problem. In large samples it could be approximately corrected by substituting estimates of the correlation coefficients in the expressions for the variances or by grouping the data.

Dependence between the samples can occur as well. This is the case if the observations in the two samples are paired through the presence of a random block effect. For example, in biological experiments where one observation is before and the other is after treatment on a patient or animal there is almost always a substantial effect due to patient or animal variability. Pairing through other block effects (viz., time, technician, litter, etc.) occurs as well. The solution for pairing is always simple. Taking the differences between the paired observations eliminates the block effects and reduces the problem to a one sample comparison of the mean difference with zero.

Other types of dependence besides pairing could occur between the samples. If the blocks contain more than just one observation from each population in each block, then the analysis is forced into a higher-way classification. More complex types of intersample dependence must be handled on an individual basis. It is the responsibility of the statistician to cross-examine the experimenter for the possible presence of any factors that might cause dependence between (or within) the samples.

For theoretical work on the effects of various types of dependence on the two sample Wilcoxon rank test see Serfling (1968) and

Hollander et al. (1974).

Exercises.

1. Show that the two sample normal theory likelihood ratio test of $H_0 : \mu_1 = \mu_2$ versus $H_1 : \mu_1 \neq \mu_2$ with $\sigma_1^2 = \sigma_2^2$ is equivalent to the two-sided two sample t test.

2. Show that for independently, identically, continuously distributed y_{ij}, $i = 1, 2$, $j = 1, \cdots, n_i$, the Mann-Whitney U statistic has variance
$$\frac{n_1 n_2 (n_1 + n_2 + 1)}{12}.$$
Hint: Prove and use the fact that
$$E[I(y_{1i} > y_{2j})\, I(y_{1i} > y_{2k})] = \frac{1}{3}$$
for $j \neq k$, where $I(\cdot)$ is the indicator function (2.16).

3. Prove that for Welch's t' test
$$\min\{n_1 - 1, n_2 - 1\} \leq \hat{\nu} \leq n_1 + n_2 - 2,$$
where the approximate degrees of freedom $\hat{\nu}$ are given by (2.44).

4. Note: This exercise will be more understandable after reading about random effects in Chapters 3 and 4.

   An investigator wants to compare Treatments A and B. On n_1 subjects paired values for Treatments A and B (i.e., one for each treatment) are available. On n_2 different subjects only the value for Treatment A is available, and on another n_3 subjects only the Treatment B is available. There is assumed to be random variation between subjects (i.e., there is a random effect a_i for subject i) as well as random variation in the paired values within subject i (i.e., there is error e_{ij} for the jth observation on subject i). Normality of random effects and errors and equal error variances should be assumed.*

Construct a Welch t'-type statistic (with approximate distribution theory) for testing the hypothesis of no difference between Treatments A and B.

5. In a Stanford Medical Center study to investigate the effectiveness of streptokinase in dissolving blood clots in the heart, many different blood values were measured including the following partial thromboplastin times (PTT) on patients who were recanalized (i.e, the clot dissolved; R) and on those who were not recanalized (NR).**

   R : 41 86 90 74 146 57 62 78 55 105 46 94 26 101 72 119 88
   NR : 34 23 36 25 35 23 87 48

   (a) Run a t test for the hypothesis of no difference in PTT for those patients who were recanalized versus those who were not.
   (b) Run a t test on the square root transforms of the data.
   (c) Run a Welch's t' test.
   (d) Run a median χ^2 test.
   (e) Run a Wilcoxon rank test.
   (f) Which test(s) do you consider most appropriate?

6. In a study of cellular immunity in infectious mononucleosis, two groups of healthy controls were considered. One group consisted of 16 Epstein-Barr virus (EBV) seropositive donors, and

---

* See Ekbohm, G. (1976), On comparing means in the paired case with incomplete data responses, *Biometrika* **63**, 299–304, for the general problem.

** Alderman, E. L., Jutzy, K. R., Berte, L. E., Miller, R. G., Friedman, J. P., Creger, W. P., and Eliastam, M. (1984). Randomized comparison of intravenous versus intracoronary streptokinase for myocardial infarction. *American Journal of Cardiology* **54**, 14–19.

the other of 10 EBV seronegative donors. These two groups were compared for lymphocyte blastogenesis with phytohemagglutinin and several EBV and control antigens.* The following stimulation indices are with the P3HR-1 virus concentrate as antigen.

Seropositive :	2.9	12.1	2.6	2.5	2.8	15.8	3.2	1.8
	7.8	2.9	3.2	8.0	1.5	6.3	1.2	3.5
Seronegative :	4.5	1.3	1.0	1.0	1.3	1.9	1.3	2.1
	2.1	1.0						

Select what you consider to be an appropriate two sample test, and test for no difference between seropositive and seronegative donors with regard to P3HR-1 concentrate.

* Nikoskelainen, J., Ablashi, D. V., Isenberg, R. A., Neel, E. U., Miller, R. G., and Stevens, D. A. (1978). Cellular immunity in infectious mononucleosis. II. Specific reactivity to Epstein-Barr virus antigens and correlation with clinical and haemotologic parameters. *Journal of Immunology* **121**, 1239–1244.

Chapter 3

ONE-WAY CLASSIFICATION

In discussing problems that involve more than two populations one may as well consider the general case of I populations because the ideas and methods are the same whether there are three, four, or more populations. The data now consist of a double array $\{y_{ij}\}$ of observations where y_{ij} denotes the jth observation in the sample from the ith population.

The model customarily chosen for data in a one-way classification is

$$y_{ij} = \mu + \alpha_i + e_{ij}, \tag{3.1}$$

where μ denotes a general overall mean, $\mu_i = \mu + \alpha_i$ denotes the mean of the ith population, and e_{ij} is random (unexplained) variation. An important distinction in the model assumptions and associated analyses arises over whether the conclusions from the statistical analysis are to apply strictly to the I populations in the experiment or whether they are to apply to a wider class of populations of which the I populations are a representative subset. In the first instance the I populations are viewed as *fixed*, whereas in the second they are considered *random*.

To illustrate this point consider an experiment comparing the effects of three drugs, each of which is a new compound developed by the laboratory. Information is desired on the comparative effects of these three agents, and there are no other compounds of interest at the moment. In this case the three populations would be assumed fixed. Similar sets of variables that are usually considered fixed are treatment regimens, types of disease, sex, age groupings, etc. In each of these cases the populations included in the experiment comprise the entire spectrum of possible populations of interest or at least most of the spectrum.

On the other hand, variables that are usually considered random are people, animals, days, etc. This is because the ones selected for the experiment are not so important in themselves. They serve instead as representatives of the whole class of all people, all animals, and all days. Conclusions based on them will be applied to the whole class.

How a variable should be treated (i.e., fixed or random) depends on how wide the inference is to be. Consider an experiment comparing the measurements made by five different lab technicians. If the five are the only five employed in the laboratory and the comparability of their results is all that matters to the lab director, then the five populations (i.e., technicians) should be assumed fixed. If, on the other hand, the five were selected to investigate the consistency between technicians in general in performing these measurements, then the inference extends beyond just these five and they should be considered random.

It is often the case in experimental work that people, animals, days, etc., are not actually selected randomly from a larger population. They are what become available to the investigator at the time of the experiment. Usually it is safe to assume their availability is the result of a process that is sufficiently haphazard to assure that no bias is involved. However, if their representativeness is in question, then they cannot be used for the estimation of the class characteristics in the fashion described in this chapter.

This chapter is divided into separate subchapters depending on whether the population effects are assumed to be fixed or random.

FIXED EFFECTS

3.1. Normal Theory.

The complete model is

$$y_{ij} = \mu_i + e_{ij}, \tag{3.2a}$$

or

$$y_{ij} = \mu + \alpha_i + e_{ij}, \tag{3.2b}$$

for $i = 1, \cdots, I$, $j = 1, \cdots, n_i$, where the e_{ij} are independently distributed as $N(0, \sigma^2)$. To avoid identifiability problems the parameters α_i are constrained by $\sum_{i=1}^{I} n_i \alpha_i = 0$.* In a balanced design with equal sample sizes the subscript i is dropped from n. The general statistical task is to construct point and interval estimates for the μ_i, or μ and α_i, or to test hypotheses about the μ_i or α_i.

3.1.1. Analysis of Variance (ANOVA)

The likelihood ratio approach leads to the standard analysis of variance displayed in Table 3.1.

Often the sum of squares for populations in Table 3.1 is computed in the form $\left(\sum_{i=1}^{I} n_i\, \bar{y}_{i\cdot}^2\right) - N\bar{y}_{\cdot\cdot}^2$ and the error sum of squares obtained by subtraction.

The mean sum of squares or mean squares (*MS*) for any effect is the effect's sum of squares (*SS*) divided by its degrees of freedom (*df*), i.e., $MS = SS/df$. Most packaged computer programs print out the *MS* column to the right of the columns in Table 3.1 and give the F ratio as well. The mean squares for error

$$\hat{\sigma}^2 = MS(E) = \frac{1}{N-I} \sum_{i=1}^{I} \sum_{j=1}^{n_i} (y_{ij} - \bar{y}_{i\cdot})^2 \tag{3.3}$$

---

* Another constraint sometimes used is $\sum_{i=1}^{I} \alpha_i = 0$. In the balanced design the two constraints are the same.

is the ANOVA estimate of the variance of the underlying normal distributions. It is the generalization to I populations of (2.2) since it pools the variability estimates from within each of the I populations.

Table 3.1. ANOVA Table [a]

VDT	df	SS
Mean (M)	1	$N\,\bar{y}_{\cdot\cdot}^2$
Populations (A)	$I-1$	$\sum_{i=1}^{I} n_i(\bar{y}_{i\cdot}-\bar{y}_{\cdot\cdot})^2$
Error (E)	$N-I$	$\sum_{i=1}^{I}\sum_{j=1}^{n_i}(y_{ij}-\bar{y}_{i\cdot})^2$
Total	N	$\sum_{i=1}^{I}\sum_{j=1}^{n_i} y_{ij}^2$

In the special case of a balanced design (i.e., $n_i \equiv n$) the expressions in Table 3.1 simplify. In particular, $SS(A) = n\sum_{i=1}^{I}(\bar{y}_{i\cdot} - \bar{y}_{\cdot\cdot})^2 = \left(n\sum_{i=1}^{I}\bar{y}_{i\cdot}^2\right) - N\,\bar{y}_{\cdot\cdot}^2$, $N = nI$, and $N - I = I(n-1)$.

The distribution theory for the sums of squares in Table 3.1 is quite simple:

$$
\begin{aligned}
SS(M) &\sim \sigma^2\,\chi_1^2\left(\frac{N\mu^2}{\sigma^2}\right),\\
SS(A) &\sim \sigma^2\,\chi_{I-1}^2\left(\frac{\sum_{i=1}^{I} n_i\alpha_i^2}{\sigma^2}\right),\\
SS(E) &\sim \sigma^2\,\chi_{N-I}^2,
\end{aligned}
\tag{3.4}
$$

[a] "VDT" abbreviates "variation due to." "df" abbreviates "degrees of freedom." "SS" abbreviates "sum of squares."
$N = \sum_{i=1}^{I} n_i$ = total sample size.
$\bar{y}_{i\cdot} = \frac{1}{n_i}\sum_{j=1}^{n_i} y_{ij}$ = sample mean of ith population.
$\bar{y}_{\cdot\cdot} = \frac{1}{N}\sum_{i=1}^{I}\sum_{j=1}^{n_i} y_{ij}$ = overall sample mean.

and the three sums of squares are independent.* The expected mean squares are

$$E(MS(M)) = \sigma^2 + N\mu^2,$$

$$E(MS(A)) = \sigma^2 + \frac{\sum_{i=1}^{I} n_i\,\alpha_i^2}{(I-1)}, \tag{3.5}$$

$$E(MS(E)) = \sigma^2.$$

The *likelihood ratio test* or *F test* of the null hypothesis $H_0 : \alpha_i \equiv 0$ vs. the very general alternative $H_1 : \alpha_i \not\equiv 0$ compares the ratio

$$F = \frac{MS(A)}{MS(E)} \tag{3.6}$$

with the percentage points of an F distribution with $(I-1)$ df in the numerator and $N - I$ df in the denominator. The upper tail of the F distribution gives the significance level. There is no differentation between one and two-tailed significance levels in the F test because of the general nature of the alternative.

3.1.2. Multiple Comparisons

The likelihood ratio test is intuitive because of the $E(MS)$ in (3.5). Under the alternative hypothesis, the F ratio tends to have larger values than if it had a central F distribution. Although numerous optimality properties have been established for the F test, it has several deficiencies.

The first is that if you conclude the population means are not all equal, the test does not tell you which means differ from which other ones. This deficiency motivated the development of multiple comparisons, which was pioneered by John Tukey and Henry Scheffé.

---

* "$\chi^2_\nu(\delta^2)$" denotes a χ^2 distribution (or variable) with ν degrees of freedom and noncentrality parameter δ^2, that is, the distribution of $\sum_{i=1}^{\nu} y_i^2$, where the y_i are independently distributed as $N(\mu_i, \sigma^2)$, $i = 1, \cdots, \nu$, and $\delta^2 = \sum_{i=1}^{\nu} \mu_i^2/\sigma^2$.

A treatise on the work in multiple comparisons is given by Miller (1981). For a shorter synopsis see Miller (1985).

If the sample sizes are equal (i.e., $n_i \equiv n$), the procedure I would use in preference to the F test is the *Tukey studentized range test*. It hinges on the probability statement

$$P\{\mu_i - \mu_{i'} \in \bar{y}_{i\cdot} - \bar{y}_{i'\cdot} \pm q^{\alpha}_{I,I(n-1)} \frac{\hat{\sigma}}{\sqrt{n}}, \quad \text{for all} \quad i, i'\} = 1 - \alpha, \tag{3.7}$$

where $q^{\alpha}_{I,I(n-1)}$ is the upper 100α percentile of the studentized range distribution with I variables entering the numerator range and $I(n-1)$ df for the error standard deviation in the denominator.* Good tables of the studentized range appear in Harter (1960, 1969a), Miller (1981), Owen (1962), and Pearson and Hartley (1970).

The overall significance of the differences in the I means is the probability that a studentized range variable $q_{I,I(n-1)}$ exceeds the observed value $\max_{i,i'}\{\sqrt{n}|\bar{y}_{i\cdot} - \bar{y}_{i'\cdot}|/\hat{\sigma}\}$. Ordinarily, significance of an individual difference $\bar{y}_{i\cdot} - \bar{y}_{i'\cdot}$ would be assessed by calculating the P value of $\sqrt{(n/2)}\,|\bar{y}_{i\cdot} - \bar{y}_{i'\cdot}|/\hat{\sigma}$ from the upper tail of a t distribution with $I(n-1)$ df. However, the most extreme difference $\max_i\{\bar{y}_{i\cdot}\} - \min_i\{\bar{y}_{i\cdot}\}$ necessarily tends to be larger than the difference between two sample means because of the selection of the largest and smallest means out of the set. Allowance for the multiple comparisons is made by using the studentized range distribution instead of the t distribution to evaluate the statistical significance of any individual difference.

Confidence intervals for each of the $\binom{I}{2}$ mean differences are given by the intervals inside the probabiity sign in (3.7). Treated

---

* A studentized range variable $q_{k,\nu}$ is distributed as $\max_{i,i'=1,\cdots,k}\{|y_i - y_{i'}|\} /(\chi^2_\nu/\nu)^{1/2}$, where $y_1, \cdots, y_k$ are independent $N(0,1)$, χ^2_ν has a χ^2 distribution with ν df, and χ^2_ν and $y_1, \cdots, y_k$ are independent.

individually, the mean differences would have confidence intervals

$$\mu_i - \mu_{i'} \in \bar{y}_{i\cdot} - \bar{y}_{i'\cdot} \pm t_{I(n-1)}^{\alpha/2} \hat{\sigma} \sqrt{\frac{2}{n}}, \tag{3.8}$$

but since there are a number of such intervals, the probability of all of them being correct is less than $1-\alpha$. This latter probability can be appreciably less than $1 - \alpha$ even for moderate values of I. Switching from multiplying $\hat{\sigma}/\sqrt{n}$ by $\sqrt{2}\ t_{I(n-1)}^{\alpha/2}$ to multiplying by $q_{I,I(n-1)}^{\alpha}$ increases the length of the intervals, but makes the probability of all the intervals being simultaneously correct equal $1 - \alpha$. In choosing whether to use the intervals (3.7) or (3.8), the statistician needs to decide whether it is the error rate on individual mean comparisons that is important to the investigation or whether it is the correctness of the whole group that is paramount.

If the design is unbalanced (i.e., $n_i \not\equiv n$), the approximate *Tukey-Kramer intervals* are available. With probability approximately $1 - \alpha$

$$\mu_i - \mu_{i'} \in \bar{y}_{i\cdot} - \bar{y}_{i'\cdot} \pm q_{I,N-I}^{\alpha}\ \hat{\sigma} \left[\frac{1}{2}\left(\frac{1}{n_i} + \frac{1}{n_{i'}}\right)\right]^{1/2} \quad \text{for all} \quad i, i'. \tag{3.9}$$

The quantity inside the square root bracket in (3.9) can be interpreted either as the sum of the sample size reciprocals for the variance of a mean difference corrected by the factor $1/2$ to convert to the studentized range, or as the harmonic mean of n_i and $n_{i'}$ inserted for n in (3.7). These intervals were originally proposed by Tukey (1953) and Kramer (1956), but they have not been used extensively because no proof existed that their probability coverage is approximately $1 - \alpha$. However, Dunnett (1980a) has shown this to be true through Monte Carlo work, and recently Hayter (1984) has proved that the probability coverage is in fact always conservative (i.e., $\geq 1 - \alpha$). Earlier Kurtz (1956) had established this for the case $I = 3$ and L. D. Brown in an unpublished 1979 proof for $I = 3$, 4, and 5.

Alternative conservative procedures have been proposed by Hochberg (1974) based on the studentized maximum modulus and by Spjøtvoll and Stoline (1973) based on the studentized augmented range. However, these confidence intervals are always broader than the Tukey-Kramer intervals. Gabriel (1978) has proposed an almost conservative procedure based on combining separate confidence intervals.

Scheffé (1953) gave an important interpretation of the F statistic that for balanced or unbalanced designs leads to the probability statement

$$P\{\mu_i - \mu_{i'} \in \bar{y}_{i\cdot} - \bar{y}_{i'\cdot} \pm ((I-1)\, F^{\alpha}_{I-1,N-I})^{1/2} \times \hat{\sigma}\left(\frac{1}{n_i} + \frac{1}{n_{i'}}\right)^{1/2}, \quad \text{for all} \quad i, i'\} \geq 1 - \alpha, \tag{3.10}$$

where $F^{\alpha}_{I-1,N-I}$ is the upper 100α percentile of the F distribution with $I-1$ df in the numerator and $N-I$ in the denominator. The simultaneous confidence intervals in (3.10) are obtained by projecting the F statistic confidence ellipsoid onto the coordinate axes for $\mu_i - \mu_{i'}$. For a balanced design the Tukey studentized range intervals given in (3.7), and for an unbalanced design, the Tukey-Kramer intervals (3.9) are shorter than the *Scheffé intervals* given in (3.10).

The *Bonferroni intervals*

$$\mu_i - \mu_{i'} \in \bar{y}_{i\cdot} - \bar{y}_{i'\cdot} \pm t^{\alpha/2K}_{N-I}\, \hat{\sigma}\left(\frac{1}{n_i} + \frac{1}{n_{i'}}\right)^{1/2} \tag{3.11}$$

also apply to balanced or unbalanced designs and are surprisingly good if K is not too large. The constant K in the probability $\alpha/2K$ for which the upper t percentage point is required is the number of confidence intervals being computed. In the one-way classification this is usually $K = \binom{I}{2}$, but it could be less if some mean comparisons are a priori not of interest. The justification for all K intervals being simultaneously correct comes from the Bonferroni inequality in elementary probability:

$$P\{A_1 \cap A_2 \cap \cdots \cap A_K\} \geq 1 - \sum_{i=1}^{K} P\{A_i^c\}, \qquad (3.12)$$

where A_i^c denotes the complement of A_i.

Special percentage points of the t distribution are required in order to use Bonferroni intervals. Tables are available in Dunn (1961) and Miller (1981) and charts in Moses (1978). A number of programmable electronic calculators have routines for calculating t percentage points and of course computers do as well.

Both the Tukey (3.7) and Scheffé's (3.10) probability statements also include confidence intervals on all possible contrasts without changing the overall probability $1 - \alpha$. A *contrast* is any linear combination of the population means $\sum_{i=1}^{I} c_i \mu_i$ for which $\sum_{i=1}^{I} c_i = 0$. Mean differences (viz., $\mu_i - \mu_{i'}$) are contrasts, and they are the parametric comparisons customarily of interest in data analysis. On occasion, however, the populations may subdivide into groups having similar characteristics (defined independently of the data) in which case comparisons of group averages such as

$$\frac{\mu_1 + \mu_2}{2} - \frac{\mu_3 + \mu_4 + \mu_5}{3} \qquad (3.13)$$

may also be of interest, and these too are contrasts.

For a balanced design the Tukey intervals for contrasts are

$$\sum_{i=1}^{I} c_i \mu_i \in \sum_{i=1}^{I} c_i \bar{y}_{i\cdot} \pm q^{\alpha}_{I,I(n-1)} \frac{\hat{\sigma}}{\sqrt{n}} \frac{1}{2} \sum_{i=1}^{I} |c_i|, \qquad (3.14)$$

and the Scheffé intervals for balanced or unbalanced designs are

$$\sum_{i=1}^{I} c_i \mu_i \in \sum_{i=1}^{I} c_i \bar{y}_{i\cdot} \pm ((I-1)\, F^{\alpha}_{I-1,N-I})^{1/2}\, \hat{\sigma} \left(\sum_{i=1}^{I} \frac{c_i^2}{n_i} \right)^{1/2} \qquad (3.15)$$

The probability that all the intervals in (3.14) [or (3.15)] are simultaneously correct for all possible contrasts is $1-\alpha$. Although the Tukey intervals are shorter than the Scheffé intervals for mean differences, the Scheffé intervals can be shorter for other contrasts like (3.13).

If the number of contrasts of interest is small, the Bonferroni intervals

$$\sum_{i=1}^{I} c_i \mu_i \in \sum_{i=1}^{I} c_i \bar{y}_{i\cdot} \pm t_{N-I}^{\alpha/2K} \hat{\sigma} \left(\sum_{i=1}^{I} \frac{c_i^2}{n_i} \right)^{1/2}, \tag{3.16}$$

where K is the total number of mean differences and contrasts of interest, may be competitive in length to (3.14) and (3.15).

3.1.3. Monotone Alternatives

A second deficiency of the F test is that it has uniform power against alternatives in all possible directions. The power is constant for all alternatives $(\mu_1, \cdots, \mu_I)$ that yield the same noncentrality parameter $\delta^2 = \sum_{i=1}^{I} n_i(\mu_i - \bar{\mu})^2/\sigma^2$ where $\bar{\mu} = \sum_{i=1}^{I} n_i\mu_i/N$. Therefore, it cannot be especially sensitive to alternatives in any particular direction.

If there is auxiliary information available in the experiment about the direction in which the alternative might lie, then it is more sensible to use a specially designed test with increased power in that direction. The all-purpose F and studentized range tests cannot win in competition with a test against a special alternative when, in fact, the special alternative is true. Of course, if the special alternative is incorrectly selected and a different, far removed alternative holds true, then the special test will fail miserably.

A case in point involves *monotone alternatives*. It may be known that if $\mu_1 = \mu_2 = \cdots = \mu_I$ does not hold, then $\mu_1 \leq \mu_2 \leq \cdots \leq \mu_I$ (with strict inequality at some point) does hold.* This

* The original subscripts labeling the populations might have to be changed

could be the case, for example, if the sequence of populations is determined by an increasing sequence of dosage levels of a drug or by staging of disease severity (e.g., stages I to IV of Hodgkin's disease). If the auxiliary information is in quantitative form with a value x_i such as dosage level associated with population i, then it is appropriate to apply regression analysis, which is discussed in Chapter 5. If, however, the extra information is qualitative as with staging of disease, then an analysis appropriate to a general monotone alternative should be employed to increase the power of detecting an increase in the means.

Bartholomew (1959a,b, 1961a,b) developed the likelihood ratio approach to monotone alternatives, and earlier Brunk (1955, 1958) had studied the associated estimation problem. Maximum likelihood estimates of the mean parameters can be derived under the restriction $\mu_1 \leq \cdots \leq \mu_I$ on the parameter space. These estimates are computed by taking a (weighted) average of any successive pair of sample means that are not in the correct monotone order. This process is continued until a monotonic sequence of sample means is obtained. The order in which the averaging process is performed is immaterial because the end result is always the same monotonic sequence.

Although the maximum likelihood estimates are easily calculated, the corresponding likelihood ratio test has a complicated null distribution that necessitates the computation of special tables. Tables have been produced for balanced designs (see Chacko, 1963; Barlow et al., 1972; or Nelson, 1977), but the unbalanced case remains hopeless. Also, the behavior of this test under alternative hypotheses and under departures from assumptions has not been studied extensively. For a summary of what is known in this area the reader is

---

to produce this ordering, but as long as the change is dictated by auxiliary a priori information and not the data, the change is okay.

referred to the excellent treatise on this approach by Barlow et al. (1972).

Because of the disadvantages associated with the maximum likelihood approach I am inclined to use a second approach due to Abelson and Tukey (1963), which is very easy computationally and does not require special tables. The power of this second test is good and, in general, its properties are more obvious.

Abelson and Tukey advocated selection of a contrast that would be sensitive to the type of alternatives considered likely. The t statistic associated with the contrast $\mathbf{c} = (c_1, \cdots, c_I)$ is

$$t = \frac{\sum_{i=1}^{I} c_i \bar{y}_{i\cdot}}{\hat{\sigma}\sqrt{\sum_{i=1}^{I} \frac{c_i^2}{n_i}}}, \tag{3.17}$$

where $\hat{\sigma}$ is given by (3.3). This statistic has a noncentral t distribution with $N - I$ df and noncentrality parameter*

$$\delta = \frac{\sum_{i=1}^{I} c_i \mu_i}{\sigma\left(\sum_{i=1}^{I} \frac{c_i^2}{n_i}\right)^{1/2}}. \tag{3.18}$$

In the balanced case the sample size n factors out of the denominator sum, and δ^2 becomes a function of the ratio $\left(\sum_{i=1}^{I} c_i \mu_i\right)^2 / \sum_{i=1}^{I} c_i^2$. For alternatives $\boldsymbol{\mu} = (\mu_1, \cdots, \mu_I)$ on the sphere with $\sum_{i=1}^{I} (\mu_i - \bar{\mu})^2$ equal to a constant, the power of the test is a function of

$$r^2 = \frac{\left(\sum_{i=1}^{I} c_i \mu_i\right)^2}{\sum_{i=1}^{I}(\mu_i - \bar{\mu})^2 \sum_{i=1}^{I} c_i^2}, \tag{3.19}$$

which is the square of the correlation coefficient between the direction in which the test is looking (i.e., $\mathbf{c}$) and the real direction (i.e., $\boldsymbol{\mu}$).

* A noncentral $t_\nu(\delta)$ variable is distributed as $y/(\sigma^2\chi_\nu^2/\nu)^{1/2}$, where y is distributed as $N(\mu, \sigma^2)$ with $\delta = \mu/\sigma$, χ_ν^2 has a χ^2 distribution with ν df, and y and χ_ν^2 are independent.

If the direction of $\boldsymbol{\mu}$ from H_0 were known, the power would be maximized by choosing $c_i = c(\mu_i - \bar{\mu})$, $i = 1, \cdots, I$, for arbitrary $c > 0$. When $\boldsymbol{\mu}$ is unknown but hypothesized to lie in a region R, Abelson and Tukey adopt a maximin approach and recommend selecting $\mathbf{c}^*$ satisfying

$$\min_{\boldsymbol{\mu} \in R} r^2(\mathbf{c}^*, \boldsymbol{\mu}) = \max_{\mathbf{c}} \min_{\boldsymbol{\mu} \in R} r^2(\mathbf{c}, \boldsymbol{\mu}). \tag{3.20}$$

They discuss the geometry of finding the maximin contrast which for convex R lies on a boundary. For monotone alternatives the region is $R = \{\boldsymbol{\mu} | \mu_1 \leq \mu_2 \leq \cdots \leq \mu_I\}$, and Abelson and Tukey have tabled the maximin contrasts $\mathbf{c}^*$ for $I \leq 20$.

Having to use special tables in a journal is a nuisance, and there are other simple contrasts whose efficiency is very high. The *linear contrasts* are the sets of coefficients for estimating the slope in a regression with equally spaced values of the independent variable. The linear $c_1, \cdots, c_I$ are displayed in (3.21) for $I = 3(1)7$, where they have been normalized into integer form.

$$\begin{array}{llllllll}
I = 3: & & & -1 & 0 & +1 & & \\
I = 4: & & -3 & -1 & +1 & +3 & & \\
I = 5: & & -2 & -1 & 0 & +1 & +2 & \\
I = 6: & -5 & -3 & -1 & +1 & +3 & +5 & \\
I = 7: & -3 & -2 & -1 & 0 & +1 & +2 & +3
\end{array} \tag{3.21}$$

More weight can be assigned to the extremes in an effort to detect a slow increase. The *linear-2 contrast* doubles the weight at the end values as in (3.22) for $I = 7$.

$$I = 7: \quad -6 \quad -2 \quad -1 \quad 0 \quad +1 \quad +2 \quad +6. \tag{3.22}$$

The *linear-2-4 contrast* doubles the penultimate value and quadruples the last coefficient as in (3.23).

$$I = 7: \quad -12 \quad -4 \quad -1 \quad 0 \quad +1 \quad +4 \quad +12. \qquad (3.23)$$

Any of these contrasts are easy to remember and use in conjunction with the test statistic (3.17).

Abelson and Tukey define efficiency to be the ratio of the respective $\min r^2$. With this definition the efficiency of the linear contrast relative to the maximin contrast is 84% at $I = 5$, but it falls off rapidly for larger I. The linear-2 contrast has over 90% efficiency up to $I = 11$ and then drops to 80% at $I = 18$. The linear-2-4 maintains efficiency greater than 95% through $I = 20$.

For a discussion of contrasts to measure quadratic effects see Section 4.1.3.

3.2. Nonnormality.

3.2.1. Effect

Lack of normality has very little effect on the significance level of the F test, even less than in the two sample case.

The asymptotic robustness of the F test follows from the multivariate central limit theorem which establishes that (3.6) has an asymptotic χ^2 distribution with $I - 1$ df as $n_i \to \infty$, $i = 1, \cdots, I$, for any underlying distribution with finite variance. The robustness improves with increasing I because the central limit theorem also smooths the sum (of squares) in the numerator as $I \to \infty$.

In a series of papers by Pearson (1931), Geary (1947), Gayen (1950a), Box and Andersen (1955), and others, Monte Carlo sampling and moment calculations have been employed to further substantiate the robustness of the F test. The reader is referred to Scheffé (1959, Section 10.3) for a thorough discussion of the present state of knowledge. In particular, Scheffé's Table 10.3.2, which is

Box and Andersen's Table 2, clearly indicates the insensitivity of the significance level to γ_1^2 between 0 and 1 and γ_2 between -1 and 1 when $I = 5$, $J = 5$. The effects can be more serious, however, for a badly unbalanced experiment.

The robustness of the studentized range has not been as thoroughly studied. It may be a bit more sensitive to nonnormality than the F test because the numerator is determined by the extreme means $\max_i\{\bar{y}_{i\cdot}\}$ and $\min_i\{\bar{y}_{i\cdot}\}$. However, as long as no n_i is too small, the central limit theorem should be making the $\bar{y}_{i\cdot}$ approximately normal and the studentized range should be approximately correct. A paper in this area is R. A. Brown (1974).

The Abelson-Tukey monotonicity test should also be insensitive to nonnormality since it only needs $\sum_{i=1}^{I} c_i\, \bar{y}_{i\cdot}$ to be normally distributed. The central limit theorem and the averaging by the c_i should help achieve this. The worst situation would be where a few means dominate the contrast as in the linear-2-4 contrast. This would be further aggravated if the dominating means were based on just a few observations.

The reader should remain aware that although the significance levels for the normal theory tests are robust for validity, these tests may not be the most powerful for nonnormal distributions. That is, they are nonrobust for efficiency. Transformations to improve normality or other tests can lead to more efficient procedures for nonnormal distributions.

3.2.2. Detection

The same devices for detecting nonnormality are available to the statistician as were previously available for the one and two sample problems (see Sections 1.2.2 and 2.2.2). My recommendation would be to make I separate probit plots, one for each sample, but calculation of the skewness and kurtosis in each sample is also sensible when it is feasible to carry out the extra computations. I would certainly not use some omnibus test over all the samples such as a combined goodness-of-fit χ^2 test or a multisample Kolmogorov-Smirnov test, but separate Shapiro-Francia tests could be computed.

3.2.3. Correction

Transformations Power transformations (1.21) in general and the square root and logarithmic transformations in particular are useful for handling positive-valued random variables with heavy upper tails. For a full discussion of transformations the reader is referred to Section 1.2.3. Even though the P value from a test on the untransformed data is reasonably robust, the power of the test and accuracy of individual confidence intervals can be improved through use of a transformation.

The choice of a particular power transformation is aided in the I sample problem by the empirical association between normality and stabilized variances. How to choose a transformation to stabilize the variances between populations is discussed in Section 3.3.3. Whichever transformation is selected by the graphical method proposed there, will probably also make the samples look more normally distributed. One can check this by probit plotting the transformed data.

Nonparametric Techniques Although it is not frequently utilized, there is a *median test* for the one-way classification due to

G. W. Brown and Mood (1948) (or see Mood, 1950, pp. 398–406). As in the two sample problem (see Section 2.2.3, "Nonparametric Techniques"), compute the median m_c for the total combined sample $\{y_{ij}, i = 1, \cdots, I, j = 1, \cdots, n_i\}$. Then within each sample count the number of observations falling above and below the median. (The simplest way of handling any observations tied with the median is to discard them.) The counts can be arranged in a $2 \times I$ table as in (3.24). (Note that the totals n_i may not quite agree with the original sample sizes due to some observations being discarded for equaling the median.)

	Sample 1	2	$\cdots$	I	
$> m_c$	a_1	a_2	$\cdots$	a_I	$a = \sum_{i=1}^{I} a_i$
$< m_c$	b_1	b_2	$\cdots$	b_I	$b = \sum_{i=1}^{I} b_i$
	n_1	n_2	$\cdots$	n_I	$N = \sum_{i=1}^{I} n_i$

(3.24)

Under the null hypothesis of no differences between the I populations, the conditional distribution of $(a_1, \cdots, a_I)$ given the marginal totals is a multivariate hypergeomeric. This is too difficult to work with to obtain an exact test as in the two sample problem, but the χ^2 statistic for the equality of I proportions $\hat{p}_1 = a_1/n_1, \cdots, \hat{p}_I = a_I/n_I$ is available:

$$\begin{aligned} \chi^2 &= \frac{1}{\hat{p}(1-\hat{p})} \sum_{i=1}^{I} n_i(\hat{p}_i - \hat{p})^2, \\ &= \frac{N^2}{a(N-a)} \left(\sum_{i=1}^{I} \frac{a_i^2}{n_i} - \frac{a^2}{N} \right), \end{aligned} \tag{3.25}$$

where $\hat{p} = a/N$. Under H_0 the statistic (3.25) has an approximate

χ^2 distribution with $I-1$ df if none of the cell entries is too small.

Nemenyi (1963) has proposed an analog to the studentized range test for sign statistics (see Miller, 1981, p. 184, or 1985), but this is never used.

Cochran (1954, Sections 6.2 and 6.3) and Armitage (1955) have proposed a test for trend in binomial proportions, which for monotone alternatives could be applied to (3.24) in conjunction with the contrasts (3.21)–(3.23). This test is also described in Armitage (1971, pp. 363–365).

Of the nonparametric tests the best known and most widely used is the *Kruskal-Wallis test* (1952). It is the analog of the F test using Wilcoxon ranks. Replace each observation y_{ij} by its rank R_{ij} in the combined sample of $N = \sum_{i=1}^{I} n_i$ observations. For each population compute the average rank score $\bar{R}_{i\cdot} = \sum_{j=1}^{n_i} R_{ij}/n_i$. As all the sample sizes become large the average rank vector $(\bar{R}_{1\cdot}, \cdots, \bar{R}_{I\cdot})$ has a limiting multivariate normal distribution. Under the null hypothesis the limiting covariance matrix has the proper form for the sum of squares $\sum_{i=1}^{I} n_i(\bar{R}_{i\cdot} - \bar{R}_{\cdot\cdot})^2$ to have a limiting χ^2 distribution with $I-1$ df except for a multiplicative constant. This constant can be determined theoretically without resorting to a sample estimate of dispersion as in the denominator of the F test. The resulting statistic is

$$
\begin{aligned}
KW &= \frac{12}{N(N+1)} \sum_{i=1}^{I} n_i(\bar{R}_{i\cdot} - \bar{R}_{\cdot\cdot})^2, \\
&= \left(\frac{12}{N(N+1)} \sum_{i=1}^{I} n_i\, \bar{R}_{i\cdot}^2 \right) - 3(N+1),
\end{aligned}
\tag{3.26}
$$

whose P value can be determined from the upper tail of a χ^2 distribution with $I-1$ df if none of the sample sizes is too small. Kruskal and Wallis (1952) give some exact probabilities for $I = 3$ and $n_i \leq 5$; similar tables appear in Kraft and van Eeden (1968, Table F), Hollander and Wolfe (1973, Table A.7), and Lehmann (1975, Table I). Extended tables for $I = 3(n_i \leq 8)$, $4(n_i \leq 4)$, $5(n_i \leq 3)$ are given by Iman et al. (1975).

When ties are present, average ranks can be used (see Section 2.2.3, "Nonparametric Techniques"). If ties occur excessively, the denominator of KW can be multiplied by the correction factor $1 - [\sum_{i=1}^{m}(t_i^3 - t_i)/(N^3 - N)]$, just as in (2.20), where t_i is the number of ties at the ith distinct value.

For deciding which populations differ, Nemenyi (1963) proposed a multiple comparisons method based on Scheffé- type projections of the Kruskal-Wallis statistic (see Miller, 1981, pp. 165–172). Dunn (1964) used the same test with Bonferroni critical constants. A slightly more powerful procedure is to reject the equality of F_i and $F_{i'}$ when

$$|\bar{R}_{i\cdot} - \bar{R}_{i'\cdot}| > q_{I,\infty}^{\alpha} \left[\frac{N(N+1)}{12}\right]^{1/2} \left[\frac{1}{2}\left(\frac{1}{n_i} + \frac{1}{n_{i'}}\right)\right]^{1/2}, \qquad (3.27)$$

where $q_{I,\infty}^{\alpha}$ is the upper 100α percentile of the studentized range distribution for a range of I variables and infinite df for the standard deviation in the denominator. This test is a rank analogue of the Tukey-Kramer test (see Section 3.1.2).

A test I like just as well as the Kruskal-Wallis and Nemenyi tests is a rank analogue to the studentized range test due to Steel (1960) and Dwass (1960) (see Miller, 1981, pp. 153–157). It is based on the comparison of each pair of populations by the Wilcoxon statistic. It is easiest to describe in its Mann-Whitney form so let

$$U_{ii'} = \sum_{j=1}^{n_i} \sum_{k=1}^{n_{i'}} I\{y_{ij} > y_{i'k}\}, \qquad (3.28)$$

where

$$I\{y_{ij} > y_{i'k}\} = \begin{cases} 1 & \text{if } y_{ij} > y_{i'k}, \\ 1/2 & \text{if } y_{ij} = y_{i'k}, \\ 0 & \text{if } y_{ij} < y_{i'k}. \end{cases} \tag{3.29}$$

The 1/2 in (3.29) is the standard tie correction, but there should not be too many ties in order for the subsequent distribution theory to hold. In order to standardize for unequal sample sizes let

$$\bar{U}_{ii'} = \frac{U_{ii'}}{n_i n_{i'}}. \tag{3.30}$$

Then the *Steel-Dwass test* compares

$$|\bar{U}_{ii'} - 1/2| \tag{3.31}$$

with the asymptotic critical value

$$q^{\alpha}_{I,\infty} \left(\frac{n_i + n_{i'} + 1}{24 n_i n_{i'}} \right)^{1/2}. \tag{3.32}$$

In (3.31) the quantity 1/2 is the theoretical mean of $\bar{U}_{ii'}$ under H_0; in (3.32) $q^{\alpha}_{I,\infty}$ is the upper 100α percentile of a studentized range distribution for I variables in the numerator and infinite df in the denominator, and $(n_i + n_{i'} + 1)/12 n_i n_{i'}$ is the variance of $\bar{U}_{ii'}$ under H_0.* If (3.31) equals or exceeds (3.32) for any i, i', H_0 is rejected, and any pair of populations for which this happens is declared significantly different.

For equal sample sizes (i.e., $n_i \equiv n$) limited small sample tables for the sum of ranks distribution for $I = 3$, $n = 2(1)6$ are given in Steel (1960), and a more extensive table with $I = 2(1)10$, $n = 6(1)20(5)50, 100$ based on the asymptotic approximation for the rank critical values appears in Miller (1981).

---

* The variance is multiplied by 1/2 in (3.32) because the denominator of a studentized range consists of an estimate for the standard deviation of a numerator mean, not the standard deviation of a difference of two means.

This test has the advantage that the comparison of population i with i' is not affected by the data from the other populations as it would be in comparing $\bar{R}_{i\cdot}$ with $\bar{R}_{i'\cdot}$ in the Nemenyi-type technique (3.27). However, somewhat more ranking is required to carry out the Steel-Dwass test. Koziol and Reid (1977) have shown that the Nemenyi-type test (3.27) and the Steel-Dwass test (3.31) and (3.32) are asymptotically equivalent for sequences of alternatives tending to H_0.

Simultaneous confidence intervals for the location differences between pairs of populations can be constructed with the Steel-Dwass ranking by the graphical method described in Section 2.2.3, "Nonparametric Techniques," when the critical constant

$$\frac{n_i n_{i'}}{2} - \frac{1}{2} - q^{\alpha}_{I,\infty} \left[\frac{n_i n_{i'}(n_i + n_{i'} + 1)}{24} \right]^{1/2} \tag{3.33}$$

is substituted for $u^{\alpha/2}$ in (2.21).

The Kruskal-Wallis, Nemenyi, and Steel-Dwass tests do not utilize any prior information on the ordering of the populations (if it exists), but there is a rank test for monotone alternatives due to Jonckheere (1954), which was proposed somewhat earlier by Terpstra (1952) in a less accessible journal. Let $U_{ii'}$ be defined as in (3.28). Then the test statistic is

$$M = \sum_{i>i'} U_{ii'}. \tag{3.34}$$

The rationale being the statistic is that if $\mu_1 \leq \cdots \leq \mu_I$, then $U_{ii'}$ should be larger than its null mean for $i > i'$. Summation of the two sample Wilcoxon statistics over the $I(I-1)/2$ pairs where $i > i'$ should accumulate any stochastic tendencies for the y_{ij} to increase as i increases. The sample sizes can be unequal, and for alternatives in the direction $\mu_1 \leq \cdots \leq \mu_I$ the null hypothesis should be rejected for large values of M. The null mean is

$$\frac{1}{4}\left(N^2 - \sum_{i=1}^{I} n_i^2\right), \tag{3.35}$$

where $N = \sum_{i=1}^{I} n_i$, and the null variance is

$$\frac{1}{72}\left[N^2(2N+3) - \sum_{i=1}^{I} n_i^2(2n_i+3)\right]. \tag{3.36}$$

If none of the n_i are too small, a normal approximation to the distribution of M will suffice. For additional details the reader is referred to Jonckheere (1954) or Hollander and Wolfe (1973, pp. 120–123). For small sample tables see Hollander and Wolfe (1973, Table A.8).

Chacko (1963) has given a rank analogue to Bartholomew's test.

A rank test in the spirit of the Abelson-Tukey contrast tests would utilize a linear combination of the population rank scores as, for example, $L = \sum_{i=1}^{I} i\bar{R}_{i\cdot}$, where the $\bar{R}_{i\cdot}$ are the rank scores used in the Kruskal-Wallis test and the populations are assumed to be indexed in increasing order. The mean and variance of L are

$$\begin{aligned} E(L) &= \frac{(N+1)\,I(I+1)}{4}, \\ \operatorname{Var}(L) &= \frac{N(N+1)}{12}\left(\sum_{i=1}^{I} \frac{i^2}{n_i}\right) - \frac{(N+1)\,I^2(I+1)^2}{48}, \end{aligned} \tag{3.37}$$

where $N = \sum_{i=1}^{I} n_i$. Asymptotically, L is normally distributed so values of $(L - E(L))/[\operatorname{Var}(L)]^{1/2}$ can be compared with standard normal critical values.

Theoretically, it would be possible to perform a *permutation test* on the F ratio (3.7) by calculating its value for each of the $N!/n_1! \cdots n_I!$ different divisions of the N total observations into samples of sizes $n_1, \cdots, n_I$ and rejecting the null hypothesis if the observed ratio is one of the αN largest. Except for the minuscule sample sizes this is too laborious for actual use even with the aid of electronic behemoths. Expressions for the permutation moments provide justification for use of the normal test theory test (see Box and Andersen, 1955).

Robust Estimation Ringland (1983) examines robust multiple comparisons based on M-estimates.

3.3. Unequal Variances.

3.3.1. Effect

By far the best article about the effect of unequal varianes on the F test is Box (1954a), and the reader should refer to this. When the variances differ between populations, the numerator and denominator sums of squares in the F ratio (3.7) are distributed as weighted sums of squares of independent normal random variables. Since the weights are unequal, the distributions are not χ^2. Box develops the distribution theory for quadratic forms of this type and applies it to the one-way classification.

To get a glimpse of the effect of unequal variances on the F test, it suffices to examine the large sample case where all the n_i are large. The denominator mean sum of squares is converging to its expected value, which is

$$E\left[\frac{1}{N-I}\sum_{i=1}^{I}\sum_{j=1}^{n_i}(y_{ij}-\bar{y}_{i\cdot})^2\right] = \frac{1}{N-I}\sum_{i=1}^{I}(n_i-1)\sigma_i^2, \qquad (3.38)$$

where σ_i^2 is the variance of the observations from the ith population. Since $N - I = \sum_{i=1}^{I}(n_i - 1)$, the expectation (3.38) is a weighted average of the σ_i^2; call it $\bar{\sigma}^2$. The expectation of the numerator mean sum of squares under H_0 is

$$
\begin{aligned}
E\left[\frac{1}{I-1}\sum_{i=1}^{I} n_i(\bar{y}_{i\cdot}-\bar{y}_{\cdot\cdot})^2\right]
&= \frac{1}{I-1}\left[\sum_{i=1}^{I} n_i\, E(\bar{y}_{i\cdot}-\mu)^2 - NE(\bar{y}_{\cdot\cdot}-\mu)^2\right], \\
&= \frac{1}{I-1}\left[\sum_{i=1}^{I} n_i \frac{\sigma_i^2}{n_i} - N\frac{\sum_{i=1}^{I} n_i\,\sigma_i^2}{N^2}\right], \qquad (3.39) \\
&= \frac{1}{N(I-1)}\sum_{i=1}^{I}(N-n_i)\sigma_i^2 .
\end{aligned}
$$

The last expression in(3.39) is a different weighted average of the σ_i^2; call it $\bar{\sigma}_*^2$.

When the n_i are all equal, the two weighted averages agree (i.e., $\bar{\sigma}_*^2 = \bar{\sigma}^2$). This means the F ratio is centered near 1 as it should be. But the variance of the numerator is

$$
\frac{2\bar{\sigma}^4}{I-1}\left[1+\frac{(I-2)}{I(I-1)}\cdot\frac{\sum_{i=1}^{I}(\sigma_i^2-\bar{\sigma}^2)^2}{\bar{\sigma}^4}\right]. \qquad (3.40)
$$

Under χ^2 theory assuming equal variances, the quantity in brackets in (3.40) should be 1, but it obviously exceeds this when the σ_i^2 differ. Thus the actual variance is larger than the theoretical variance for the case of equal σ_i^2, and the upper tail of the distribution of the F ratio has more mass in it than anticipated by the χ_{I-1}^2 distribution. For an observed F ratio the actual P value is larger than the one calculated from the tables, but numerical studies indicate that the effect is not large. This conclusion is also born out in small samples (see Box, 1954a or Scheffé, 1959, Section 10.3).

When the n_i are unequal, the effects can be more serious. Suppose the large σ_i^2 happen to be associated with the large n_i. Then in $\bar{\sigma}^2$ of (3.38) the large σ_i^2 receive greater weight, whereas in $\bar{\sigma}_*^2$ of

(3.39) the small σ_i^2 receive greater weight. The expectation of the numerator mean squares is, therefore, less than the expectation of the denominator, and the center of the distribution of the F ratio is shifted below 1. The actual P value is less than the one stated from the tables. If the large σ_i^2 are associated with the small n_i, the shift goes in the opposite direction. The actual P value exceeds the reported one, and it can increase dramatically above its nominal level without too much disparity in the variances. The reader is referred to Table 4 in Box (1954a) or Table 10.4.2 in Scheffé (1959) to inspect the potential danger.

Falsely reporting significant results when the small samples have the larger variances is a serious worry. The lesson to be learned is to balance the experiment if it is at all possible, for then unequal variances (and other departures from assumptions) have the least effect.

A small study of the effect of unequal variances on the studentized range test has been published by R. A. Brown (1974). The results are similar to those cited for the F test.

For an Abelson-Tukey monotonicity test it is relatively easy to see what will happen. The variance of the contrast $\sum_{i=1}^{I} c_i \bar{y}_{i\cdot}$ in (3.17) is

$$\sum_{i=1}^{I} c_i^2 \frac{\sigma_i^2}{n_i}, \tag{3.41}$$

and the square of the denominator is converging in probability to

$$\left(\frac{1}{N-I}\sum_{i=1}^{I}(n_i - 1)\sigma_i^2\right)\left(\sum_{i=1}^{I}\frac{c_i^2}{n_i}\right). \tag{3.42}$$

Even with the n_i equal, if the large σ_i^2 occur at the ends of the range where the c_i are largest, the actual variance is larger than the normalizing one so the stated P value is too small. The linear-2 and linear-2-4 are the most sensitive to this. If the smaller sample sizes also occur at the ends, the effect is magnified. A reverse effect on the P value pertains when the large σ_i^2 occur in the middle. No numerical work on quantifying these comments has been carried out.

3.3.2. Detection

Use of a preliminary test of homogeneity of variances is not recommended. The three standard tests for equality of variances, which are based on normal theory, are those of Bartlett, Hartley, and Cochran, but each of these is extremely sensitive to departures from normality. There are robust tests, but they all involve substantial extra computation. This problem is the subject of Chapter 7.

It is best to avoid the problem of preliminarily testing variances. It is harder to decide the isssue of equality or the lack thereof than it is to corect for inequality if visual inspection suggests that this might be warranted.

3.3.3. Correction

Transformations are extremely useful in correcting unequal variances when the size of the variance is related to the size of the mean. Plot the I pairs $(\bar{y}_{i\cdot}, s_i)$, $i = 1, \cdots, I$, where

$$s_i^2 = \frac{1}{n_i - 1} \sum_{j=1}^{n_i} (y_{ij} - \bar{y}_{i\cdot})^2. \tag{3.43}$$

This is depicted in Figure 3.1. Often the s_i tend to incrase with increasing $\bar{y}_{i\cdot}$. With luck the statistician can make a guess on an approximate relationship $s \cong h(\bar{y})$ between the standard deviations and the means. In this case the asymptotic relation

$$SD(g(y)) \cong SD(y)|g'(\mu)|, \tag{3.44}$$

which was derived in Section 2.3.3, motivates trying the transforma-

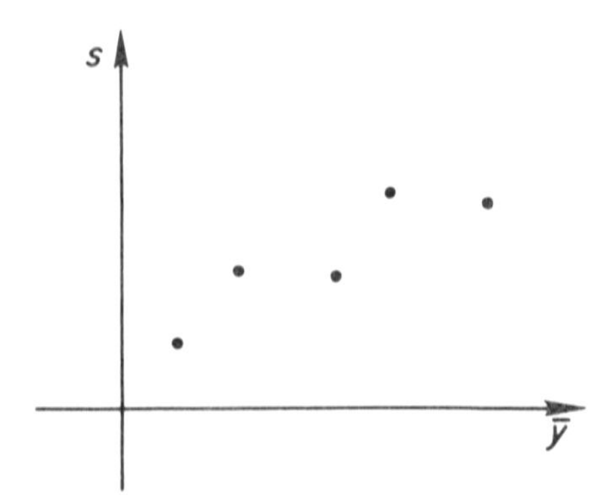

Figure 3.1

tion

$$g(y) = \int^{y} \frac{1}{h(u)} du. \tag{3.45}$$

When s increases approximately linearly with $\bar{y}$ (i.e., $s \cong a\bar{y}$), the relation (3.45) suggests trying $\log y$ or $\log(y + c)$. For a more curved relationship like $s \cong a\sqrt{\bar{y}}$, (3.45) suggests the square root transformation $\sqrt{y}$ or $\sqrt{y + c}$. Whatever transformation is selected, the prudent statistician checks the variances of the transformed data to ascertain if the transformation has in fact stabilized the variances.

Transformations are not as useful when the data can be both positive and negative, and when the variances do not have a monotonic relationship with the means. For these contingencies the alternative nonparametric tests are available. The Brown-Mood median test and the Kruskal-Wallis rank test (see Section 3.2.3) should be fairly insensitive to moderately unequal variances, but no study of this has been published to date.

Tamhane (1979) and Dunnett (1980b) compare various Welch-type (see Section 2.3.3, "Other Tests") procedures that have been proposed for the multiple comparisons problem with $\sigma_i^2 \not\equiv \sigma^2$.

For monotone alternatives one can substitute the sample estimats s_i^2 for the unknown σ_i^2, $i = 1, \cdots I$, in the variance (3.41) for $\sum_{i=1}^{I} c_i \bar{y}_{i\cdot}$. A Satterthwaite (1946) approximation could be used to approximate the degrees of freedom of this variance estimate.

3.4. Dependence.

Dependence in the data caused by blocking or grouping of the observations is easily handled. Extra parameters are added to the model (3.1) to represent the nuisance effects. The model then becomes a two-way or higher-way classification, and the appropriate analysis for these more complex designs should be applied.

The presence of serial correlation within or between the samples from the different populations is a much more serious affair. Box (1954b) studied the effects in a two-way classification, and the results are interpretable for the one-way classification as well. Serial correlation within the samples from each population badly distorts the significance level of the F test. The reported P value tends to be too large or too small depending on whether the correlation is negative or positive. The effect of serial correltaion in blocks across the populations is much less severe. For further details and numerical results the reader is referred to Box (1954b) and Scheffé (1959, Section 10.5).

The techniques available for detection of serial correlation within population samples is the same as in the one sample problem (see Section 1.3.2). It's just that there are more samples in the one-way classification. Successive pairs $(y_{ij}, y_{i,j+1})$ can be plotted for each population sample for visual inspection, or the serial correlation coefficients can be computed.

It is well to know if serial correlation is present so that it is known whether the significance level of the F test is shaky. However, if the F test is in trouble, there is precious little that can be done to rescue the situation. Estimates of the correlations can be plugged into the expressions for the mean and variance of the F ratio, and for large data sets grouping may help (see Section 1.3.3). Since the median and rank tests are also in trouble from serial dependence in the one sample problem, it is likely that they are in trouble in the one-way classification as well and cannot bail out the analysis.

RANDOM EFFECTS

3.5. Normal Theory.

The discussion here focuses on the situation where the I populations in the experiment are not the only ones of interest. They are merely representatives of a wider class of populations from which they have been selected. The experimenter and statistician are primarily seeking inferential statements about the broad class of populations.

The classical model is

$$y_{ij} = \mu + a_i + e_{ij}, \quad i = 1, \cdots, I, \quad j = 1, \cdots, n_i, \tag{3.46}$$

where the random variables a_i and e_{ij} are distributed as

$$\begin{aligned} a_i &\quad \text{independent} \quad N(0, \sigma_a^2), \\ e_{ij} &\quad \text{independent} \quad N(0, \sigma_e^2), \\ \{a_i\} &\quad \text{independent of} \quad \{e_{ij}\}. \end{aligned} \tag{3.47}$$

Whereas in the fixed effects model the analysis concentrates on estimating and testing the population differenes $\alpha_i - \alpha_{i'}$, for the random effects model estimating and testing the variances σ_a^2 and σ_e^2 are usually the primary concern. In some instances an estimate of μ or each population mean $\mu + a_i$ may also be desired.

3.5.1. Estimation of Variance Components

The classic approach is to use the expected mean squares of the ANOVA table (Table 3.1) under the random effects model and the *method of moments* to estimate σ_e^2 and σ_a^2. It is a simple computation to show that

$$E(MS(A)) = \sigma_e^2 + \frac{1}{N(I-1)}\left(N^2 - \sum_{i=1}^{I} n_i^2\right)\sigma_a^2, \qquad (3.48)$$

$$E(MS(E)) = \sigma_e^2.$$

Equating the moments to the observed mean squares and solving the pair of equations gives

$$\hat{\sigma}_e^2 = MS(E),$$

$$\hat{\sigma}_a^2 = \frac{N(I-1)\,[MS(A) - MS(E)]}{N^2 - \sum_{i=1}^{I} n_i^2}. \qquad (3.49)$$

For a balanced design (i.e., $n_i \equiv n$) the expressions for the estimates simplify to

$$\hat{\sigma}_e^2 = MS(E),$$

$$\hat{\sigma}_a^2 = \frac{MS(A) - MS(E)}{n}. \qquad (3.50)$$

With the modification that if $\hat{\sigma}_a^2$ is negative it is replaced by zero, these estimators are the ones most commonly used in practice.

For a balanced design the estimators (3.50) possess certain optimality properties. The vector $[\bar{y}_{..}, SS(E), SS(A)]$ is a complete, minimal sufficient statistic for $(\mu, \sigma_e^2, \sigma_a^2)$. The estimators (3.50) are therefore the uniform minimum variance unbiased estimators. Without the assumption of a normal distribution they are the uniform minimum variance quadratic unbiased estimators (e.g., see Graybill, 1976, pp. 614–615 or Searle, 1971, pp. 405–406). Still there are biased estimators that have more desirable properties from the point of view of mean squared error loss [i.e., $E(\hat{\sigma}_e^2 - \sigma_e^2)^2$]. These alternative estimators deserve to receive greater attention in applications.

The *Hodges-Lehmann* (1951) *estimator*

$$\hat{\sigma}_e^2 = \frac{SS(E)}{I(n-1)+2} \tag{3.51}$$

has the smallest mean squared error loss in the class of estimators of the form $c \times SS(E)$ where c is a constant.

The *maximum likelihood estimators* of σ_e^2, σ_a^2 are

$$\begin{aligned} \hat{\sigma}_e^2 &= MS(E), \\ \hat{\sigma}_a^2 &= \frac{1}{n}\left[\left(1-\frac{1}{I}\right)MS(A) - MS(E)\right], \end{aligned} \tag{3.52}$$

if $(1-I^{-1})MS(A) \geq MS(E)$, and

$$\begin{aligned} \hat{\sigma}_e^2 &= \frac{1}{n}\left[\left(1-\frac{1}{I}\right)MS(A) + (n-1)MS(E)\right], \\ \hat{\sigma}_a^2 &= 0, \end{aligned} \tag{3.53}$$

if $(1-I^{-1})MS(A) < MS(E)$. The estimate of $\hat{\sigma}_e^2$ in (3.53) is a pooled estimate of error; if $MS(A)$ is not large enough to indicate that $\sigma_a^2 > 0$, then $SS(A)$ is added to $SS(E)$ in the numerator of $\hat{\sigma}_e^2$. The expressions (3.52) and (3.53) can be written as

$$\begin{aligned} \hat{\sigma}_e^2 &= \min\left\{\frac{SS(E)}{I(n-1)}, \quad \frac{SS(E)+SS(A)}{In}\right\}, \\ \hat{\sigma}_a^2 &= \frac{1}{n}\left[\frac{SS(A)}{I} - \frac{SS(E)}{I(n-1)}\right]^+, \end{aligned} \tag{3.54}$$

where $(a)^+ = \max(a,0)$. These maximum likelihood estimators have uniformly smaller mean squared error loss than the unbiased estimators (3.50) but they in turn can be dominated by more sophisticated estimators.

The *Klotz-Milton-Zacks* (1969) *estimators*

$$\begin{aligned}
\hat{\sigma}_e^2 &= \min\left\{\frac{SS(E)}{I(n-1)+2}, \frac{SS(E)+SS(A)}{In+1}\right\}, \\
\hat{\sigma}_a^2 &= \frac{1}{n}\left[\frac{SS(A)}{I+1} - \frac{SS(E)}{I(n-1)}\right]^+,
\end{aligned} \tag{3.55}$$

improve upon (3.54) by choosing the denominator constants in a more optimal fashion.

Finally, the *Stein* (1964) *estimators*

$$\begin{aligned}
\hat{\sigma}_e^2 &= \min\left\{\frac{SS(E)}{I(n-1)+2}, \frac{SS(E)+SS(A)}{In+2}, \right. \\
&\qquad \left. \frac{SS(E)+SS(A)+SS(M)}{In+2}\right\}, \\
\hat{\sigma}_a^2 &= \min\left\{\frac{1}{n}\left(\frac{SS(A)}{I+1} - \frac{SS(E)}{I(n-1)}\right)^+, \right. \\
&\qquad \left. \frac{1}{n}\left(\frac{SS(A)+SS(M)}{I+2} - \frac{SS(E)}{I(n-1)}\right)^+\right\},
\end{aligned} \tag{3.56}$$

improve upon (3.55) by including the one df variation in the grand mean if it does not differ much from zero.* Since only one df is involved, the amount of improvement is apt to be only slight.

In their excellent paper Klotz et al. (1969) give proofs of the preceding statements on mean squared error loss and numerical comparisons of the estimators. They also consider some formal Bayes estimators. In a later paper Portnoy (1971) considers formal Bayes estimators in greater depth.

C. R. Rao (1970, 1971, 1972) introduced the concept of minimum norm quadratic unbiased estimators (MINQUE).

---

* Selection of zero is arbitrary. Any other value μ_0 that has some justification independent of the data can be used. In this event $SS(M)$ is replaced by $In(\bar{y}_{\cdot\cdot} - \mu_0)^2$.

For work on nonnegative unbiased estimators, see LaMotte (1973) and Pukelsheim (1981).

Alternative estimators for the unbalanced design (i.e., $n_i \not\equiv n$) are not considered here. The unbalanced case is more complex than the balanced. The results that exist are considerably messier and go beyond the intent of this book. The reader is referred to Chapters 10 and 11 of Searle (1971).

3.5.2. Tests for Variance Components

For a balanced design

$$\begin{aligned} SS(M) &\sim (\sigma_e^2 + n\sigma_a^2)\, \chi_1^2\left(\frac{In\mu^2}{\sigma_e^2 + n\sigma_a^2}\right), \\ SS(A) &\sim (\sigma_e^2 + n\sigma_a^2)\, \chi_{I-1}^2, \\ SS(E) &\sim \sigma_e^2\, \chi_{I(n-1)}^2, \end{aligned} \tag{3.57}$$

under the normal theory assumption, and the three sums of squares are independent.

A common problem is that of testing for the presence of population variability, that is, $H_0 : \sigma_a^2 = 0$ vs. $H_1 : \sigma_a^2 > 0$. A test that is uniformly most powerful similar and invariant and is almost the likelihood ratio test is to reject H_0 for large values of $MS(A)/MS(E)$ (see Herbach, 1959). This is the same ratio as (3.6), and under H_0 this ratio has an F distribution with df $I-1$ and $I(n-1)$ for numerator and denominator, respectively. Under the alternative H_1, the distribution is not a noncentral F as in the fixed effects model. Instead, from (3.57) it is distributed as*

$$\left(1 + \frac{n\sigma_a^2}{\sigma_e^2}\right) F_{I-1,I(n-1)}. \tag{3.58}$$

* "F_{ν_1,ν_2}" denotes an F variable (or distribution) with ν_1 and ν_2 df for numerator and denominator, respectively.

The parameter σ_a^2 enters through a multiplicative factor to a central F variable rather than through a noncentrality parameter. Power calculations are thus obtained from tables of the central F or beta distributions.

Spjøtvoll (1967) studies the structure of optimum tests of $H_0 : \sigma_a^2/\sigma_e^2 \leq \Delta_0$.

Although their general use is not recommended because of their extreme sensitivity to nonnormality (see Section 3.6.2), confidence intervals can be constructed based on the distribution theory (3.57). In particular, a confidence interval for σ_e^2 can be obtained from

$$\frac{SS(E)}{\sigma_e^2} \sim \chi^2_{I(n-1)}, \tag{3.59}$$

and a confidence interval for σ_a^2/σ_e^2 from

$$\frac{MS(A)}{MS(E)} \times \frac{\sigma_e^2}{\sigma_e^2 + n\sigma_a^2} \sim F_{I-1,I(n-1)}. \tag{3.60}$$

The ratio σ_a^2/σ_e^2 measures the size of the population variability relative to the error variability inherent in the data. In some problems this ratio may be the parameter of interest, but in others a confidence interval on σ_a^2 alone may be required. This is a much more difficult problem. Bulmer (1957) has a complicated method for constructing an approximate confidence interval, and Scheffé (1959, pp. 231–235) discusses this approach in detail. Another method that should yield a rougher approximation is to employ a Satterthwaite approximation (see Section 2.3.3, "Other Tests"):

$$\frac{1}{n}(MS(A) - MS(E)) \approx \sigma_a^2 \frac{\chi^2_\nu}{\nu}. \tag{3.61}$$

The degrees of freedom ν in the approximation are selected by equating the second moments of the random variables on the two sides of

(3.61). This yields

$$\hat{\nu} = \frac{n^2\hat{\sigma}_a^4}{\frac{(\hat{\sigma}_e^2+n\hat{\sigma}_a^2)^2}{I-1} + \frac{\hat{\sigma}_e^4}{I(n-1)}}. \tag{3.62}$$

No confidence intervals linked to the more sophisticated estimators (3.54) – (3.56) have been developed.

In the unbalanced design

$$SS(A) \sim \sigma_e^2\, \chi_{I-1}^2 \tag{3.63}$$

under $H_0 : \sigma_a^2 = 0$. Under $H_1 : \sigma_a^2 > 0$, $SS(A)$ does not have a χ^2 distribution but instead a weighted combination of χ^2 distributions. The ratio $MS(A)/MS(E)$ can still be used to test $H_0 : \sigma_a^2 = 0$ because under H_0 it has an $F_{I-1,N-I}$ distribution, but the distribution under the alternative is more complicated than in the balanced case.

A normal theory confidence interval for σ_e^2 based on

$$\frac{SS(E)}{\sigma_e^2} \sim \chi_{N-I}^2 \tag{3.64}$$

is available, but no confidence intervals have been developed for σ_a^2 or σ_a^2/σ_e^2 in the unbalanced case.

3.5.3. Estimation of Individual Effects

In most cases the primary statistical problem in a one-way classification with random effects is to estimate or test hypotheses about the two sources of variability, namely, error (σ_e^2) and populations (σ_a^2). Occasionally, one wants to estimate or test μ, and also at times to estimate the individual population means $\mu_i = \mu + a_i$, $i = 1, \cdots, I$. The latter problem is relevant when specific actions or calculations are to be made for each individual population on the basis of its estimated mean value. For examples of this the reader is referred to Efron and Morris (1975).

The standard maximum likelihood approach would be to estimate $\mu_i = \mu + a_i$ by $\bar{y}_{i\cdot}$. However, if for I estimators $\hat{\mu}_i$ of the I parameters μ_i the criterion of performance is the sum of the squared error losses $\sum_{i=1}^{I}(\hat{\mu}_i - \mu_i)^2$, then *empirical Bayes* estimators do better.

Under the distribution structure (3.47), the Bayes estimator of μ_i for the balanced design is

$$\begin{aligned}\hat{\mu}_i &= \mu + \left(1 - \frac{\sigma_e^2}{\sigma_e^2 + n\sigma_a^2}\right)(\bar{y}_{i\cdot} - \mu), \\ &= \left(\frac{\sigma_e^2}{\sigma_e^2 + n\sigma_a^2}\right)\mu + \left(\frac{n\sigma_a^2}{\sigma_e^2 + n\sigma_a^2}\right)\bar{y}_{i\cdot}\ .\end{aligned} \tag{3.65}$$

Whereas the risk of the set of maximum likelihood estimators $\{\bar{y}_{i\cdot}\}$ equals $I\sigma_e^2/n$, the Bayes risk for the set of Bayes estimators (3.65) is

$$\frac{I\sigma_e^2}{n}\left(1 - \frac{\sigma_e^2}{\sigma_e^2 + n\sigma_a^2}\right). \tag{3.66}$$

The Bayes estimator (3.65) corrects the population sample mean toward the theoretical overall mean by an amount proportional to the size of the two components of variability σ_e^2/n and σ_a^2. The savings in risk of (3.66) over $I\sigma_e^2/n$ can be considerable for small σ_a^2 relative to σ_e^2/n.

Of course, the parameters μ, σ_e^2, σ_a^2 are unknown in any practical problem unless there is previous data or auxiliary information available. However, they can be estimated from the data and this leads to empirical Bayes estimators. James and Stein (1961) showed that, in the case of known μ and σ_e^2/n but unknown σ_a^2, the estimators

$$\hat{\mu}_i = \mu + \left(1 - \frac{\sigma_e^2}{\sigma_e^2 + n\hat{\sigma}_a^2}\right)(\bar{y}_{i\cdot} - \mu), \tag{3.67}$$

where

$$\hat{\sigma}_a^2 = \frac{1}{I-2}\sum_{i=1}^{I}(\bar{y}_{i\cdot} - \mu)^2 - \frac{\sigma_e^2}{n}, \tag{3.68}$$

uniformly (over fixed a_i) improve on the risk of the maximum likelihood estimators provided $I \geq 3$. The Bayes risk (averaging over the a_i) for the estimators (3.67) is

$$\frac{I\sigma_e^2}{n}\left(1-\left(\frac{I-2}{I}\right)\frac{\sigma_e^2}{\sigma_e^2+n\sigma_a^2}\right) \tag{3.69}$$

(see Efron and Morris, 1973a). One can substitute an independent estimate of σ_e^2 based on $MS(E)$ into (3.67), and Lindley (1962) suggested substituting $\bar{y}_{..}$ as an estimate of μ. The Lindley form of the James-Stein estimator is

$$\hat{\mu}_i = \bar{y}_{..} + \left(1 - \frac{(I-3)}{I(n-1)+2}\cdot\frac{SS(E)}{SS(A)}\right)(\bar{y}_{i.} - \bar{y}_{..}), \tag{3.70}$$

which shrinks each sample mean toward the grand mean in proportion to the relative sizes of the sums of squares.

The empirical Bayes estimator (3.70) seems most relevant for application to the one-way classification (provided $I \geq 4$). There is a large literature on these estimators and variants of them which the reader may wish to pursue. He or she is referred, in particular, to James and Stein (1961) and Efron and Morris (1973). The latter authors include a discussion of unbalanced designs as well.

The original work of Stein (1956) and James and Stein (1961) was for the fixed effects model (3.2), not the random effects model (3.46). They established the existence of estimators that dominate the maximum likelihood estimators in terms of the sum of the squared error losses uniformly over all values of the mean vector $(\mu_1, \cdots, \mu_I)$ for $I \geq 3$. Lindley was one of the first to give an empirical Bayes interpretation to the James-Stein estimators. His remarks appear in the discussion following Stein (1962). Efron and Morris in their aforementioned series of articles amplify the empirical Bayes interpretation of these estimators.

Although the estimators (3.70) improve on the maximum likelihood estimators in the fixed effects case, they have not received widespread acceptance. The examples in which they have been utilized have more of a random effects flavor. It is not clear why there is hesitation in using the empirical Bayes estimators (3.70) in fixed effects problems. Perhaps there is some distrust of the loss function that adds all the squared errors. The estimation of any given mean is subjugated to the estimation of the whole set of means. Concern for the individual mean may inhibit using an estimate substantially different from the observed sample mean. Efron and Morris (1971, 1972) discuss this point and introduce modifications of the empirical Bayes estimators.

Another handicap of the estimators (3.70) is that no tests or confidence procedures are available for use in conjunction with them. Some theoretical work (viz., Stein, 1962; Joshi, 1967; Faith, 1976; and Morris, 1983) has appeared, but it has not been reduced to practical form for everyday use. The work of Dixon and Duncan (1975) is very relevant but is not entirely practical.

3.5.4. Estimation of the Overall Mean

The estimate for the overall mean μ in the balanced case is $\hat{\mu} = \bar{y}_{..}$. Its variance is

$$\frac{\sigma_e^2}{NI} + \frac{\sigma_a^2}{I}. \tag{3.71}$$

Tests and confidence intervals for μ can be constructd from $\bar{y}_{..}$ and $MS(A)/In$ since $\bar{y}_{..}$ is normally distributed and an unbiased estimate of (3.71) is $MS(A)/In$. The df for the t statistic is $I - 1$.

In the unbalanced case there are several choices of estimators for μ. One is the weighted average of the population means $\bar{y}_{..} = (1/N) \sum_{i=1}^{I} n_i \bar{y}_{i\cdot}$, and another is the unweighted average $\bar{y}^*_{..} = (1/I)$

$\sum_{i=1}^{I} \bar{y}_{i\cdot}$. The variances for these two estimators are

$$\begin{aligned} \mathrm{Var}(\bar{y}_{\cdot\cdot}) &= \frac{1}{N^2}\sum_{i=1}^{I} n_i^2\left(\frac{\sigma_e^2}{n_i}+\sigma_a^2\right), \\ \mathrm{Var}(\bar{y}_{\cdot\cdot}^{*}) &= \frac{1}{I^2}\sum_{i=1}^{I}\left(\frac{\sigma_e^2}{n_i}+\sigma_a^2\right), \end{aligned} \tag{3.72}$$

and either one can be estimated by substituting estimates for σ_e^2 and σ_a^2. The relative sizes of σ_e^2 and σ_a^2 determine which variance in (3.72) is smaller and thus which estimator is to be preferred. If $\sigma_e^2 >> \sigma_a^2$, then $\mathrm{Var}(\bar{y}_{\cdot\cdot}) < \mathrm{Var}(\bar{y}_{\cdot\cdot}^{*})$, and if $\sigma_e^2 << \sigma_a^2$, the reverse is true. By substituting estimates for σ_e^2 and σ_a^2 into (3.72) one can assess whether either estimator is definitely superior to the other.

It is possible to go even further and determine the weights $\{\hat{w}_i\}$ that give the estimator $\bar{y}_{\cdot\cdot}^{w} = \sum_{i=1}^{I} \hat{w}_i\bar{y}_{i\cdot} / \sum_{i=1}^{I} \hat{w}_i$ with the smallest estimated variance. In this case $\hat{w}_i$ is a function of $\hat{\sigma}_e^2$ and $\hat{\sigma}_a^2$. My experience has been that this approach produces a less satisfactory estimator than $\bar{y}_{\cdot\cdot}$ or $\bar{y}_{\cdot\cdot}^{*}$. The noise in the weights $\hat{w}_i$ introduced by the estimates $\hat{\sigma}_e^2$ and $\hat{\sigma}_a^2$ tends to produce an unstable result. There is some theoretical work that substantiates this assertion for small sample sizes like $n_i \leq 9$ (see Graybill and Deal, 1959, and Norwood and Hinkelmann, 1977).

3.6. Nonnormality.

3.6.1. Effect

Lack of normality can occur in both the variables a_i and the variables e_{ij}. Let $\gamma_{2,a}$ denote the kurtosis of the distribution governing the a_i, and $\gamma_{2,e}$ the kurtosis for the e_{ij}. The corresponding skewness parameters are not introduced here because their effects on the distributions of the statistics are not as profound.

Confidence intervals and tests for σ_e^2 based on the assumption

$$\frac{SS(E)}{\sigma_e^2} \sim \chi^2_{N-I} \tag{3.73}$$

are very nonrobust. This is created by the dependence of the variance of $SS(E)/\sigma_e^2$ on $\gamma_{2,e}$, which is not accounted for by the χ^2 distribution. For greater detail the reader is referred to Chapter 7.

The test of the null hypothesis $H_0 : \sigma_a^2 = 0$, which assumes that

$$\frac{MS(A)}{MS(E)} = \frac{N-I}{I-1} \cdot \frac{\sum_{i=1}^{I} n_i(\bar{y}_{i\cdot} - \bar{y}_{\cdot\cdot})^2}{\sum_{i=1}^{I}\sum_{j=1}^{n_i}(y_{ij} - \bar{y}_{i\cdot})^2} \tag{3.74}$$

is distributed as $F_{I-1,N-I}$, is quite robust. This must be the case because it coincides with the fixed effects test of $H_0 : \alpha_i \equiv 0$, which is known to be robust (see Section 3.2.1). Under H_0 the a_i are nonexistent ($\sigma_a^2 = 0$, $\gamma_{2,a} = 0$) so they do not affect the distribution of (3.74). Since the $\bar{y}_{i\cdot}$ are sample means with $\gamma_2(\bar{y}_{i\cdot}) = \gamma_{2,e}/n_i$, the effect that the $\gamma_{2,e}$ might have on the numerator is dampened so the distribution of the numerator is nearly χ^2. There are usually plenty of degrees of freedom for estimating σ_e^2 in the denominator so $\gamma_{2,e}$ does not appreciably affect the distribution of the ratio (3.74).

Under the alternative $\sigma_a^2 > 0$, the robustness vanishes when $\gamma_{2,a} \neq 0$. Unless the n_i are very small or $\sigma_e^2 >> \sigma_a^2$, the population variable a_i dominates $\bar{e}_{i\cdot}$ in controlling $\bar{y}_{i\cdot} = a_i + \bar{e}_{i\cdot}$. The effect of $\gamma_{2,a}$ on a_i has not been dampened by any averaging process and

$$\text{Var}\left(\frac{1}{I-1}\sum_{i=1}^{I}(a_i - \bar{a})^2\right) = \sigma_a^4\left(\frac{2}{I-1} + \frac{\gamma_{2,a}}{I}\right). \tag{3.75}$$

The kurtosis $\gamma_{2,a}$ thus has a substantial effect on the variance of the numerator of (3.74). Since the a_i cancel out of the denomiantor of (3.74), the denominator cannot correct for the change in the variance of the numerator. This leads to nonrobustness of the distribution of (3.74). Confidence intervals for σ_a^2/σ_e^2 or tests of the hypothesis

$H_0 : \sigma_a^2/\sigma_e^2 \leq \Delta_0$ based on (3.74) are, therefore, very sensitive to departures from normality. For numerical confirmation of this the reader is referred to Arvesen and Schmitz (1970) and Arvesen and Layard (1975).

No numerical work has appeared on the effects of nonnormality on the distribution of the alternative estimators (3.54)–(3.56) for σ_a^2. However, one would guess that $\gamma_{2,a}$ has a considerable influence.

The effect of nonnormality on the performance of the empirical Bayes estimators of $\mu_i = \mu + a_i$ is less clear. Some work in this direction would certainly aid in determining whether they should be used routinely in practice.

The effect of $\gamma_{2,e}$ and $\gamma_{2,a}$ on the estimates of the overall mean and their estimated standard errors is more straightforward, but no numerical work has been published.

3.6.2. Detection

Lack of normality of the e_{ij} is easy to spot when there are enough of them. The situation is simply an I-fold repetition of the one sample problem; therefore, the reader is referred to Section 1.2.2. Probit plots of y_{ij}, $j = 1, \cdots, n_i$, for each of the population samples should reveal any skewness or kurtosis in the error distribution.

Detection of $\gamma_{2,a} \neq 0$ is more difficult. Typically, I is not all that large so there are not many variables a_i. One can make a probit plot of the $\bar{y}_{i\cdot}$, $i = 1, \cdots, I$, but this does not allow one to see the empirical distribution of the a_i directly. Each a_i is contaminated by the addition of $\bar{e}_{i\cdot}$, which clouds the picture of the behavior of the a_i. Since tests or confidence intervals on σ_a^2 are usually more of interest than those on σ_e^2, and since $\gamma_{2,a}$ substantially affects the normal theory tests in the nonnull case, this leaves the normal theory techniques in an unfortunate situation.

3.6.3. Correction

A suitably chosen transformation may improve the normality of the data. With the use of a transformation there may be some difficulty in the interpretation of the variances on the transformed scale. It may be necessary to inversely transform the variance estimates back into the original scale (see Section 2.3.3, "Transformations").

The main alternative to normal theory for assessing the variability in variance component estimates is the jackknife. This technique is decribed below and in Chapter 7. Arvesen (1969), Arvesen and Schmitz (1970), and Arvesen and Layard (1975) have studied its application to variance component problems. Miller (1974a) has described the more general uses of the jackknife, which was proposed by Tukey (1958) for robust interval estimation.

Consider interval estimation on σ_a^2 in the balanced design. Let $\theta = \sigma_a^2$ and

$$\hat{\theta} = \frac{1}{n}(MS(A) - MS(E)). \tag{3.76}$$

The *jackknife* systematically deletes each of the I population samples in its turn and recomputes (3.76) each time with one population missing. Let $\hat{\theta}_{-k}$ be the estimate (3.76) computed from y_{ij}, $i = 1, \cdots, k-1, k+1, \cdots, I$, $j = 1, \cdots, n$. The next step is to form the quantities

$$\tilde{\theta}_k = I\hat{\theta} - (I-1)\hat{\theta}_{-k}, \quad k = 1, \cdots, I, \tag{3.77}$$

which have been called "pseudo-values" by Tukey. Then $\tilde{\theta}_1, \cdots, \tilde{\theta}_I$ are to be treated as approximately independently, identically distributed random variables so that

$$\frac{\tilde{\theta} - \theta}{\sqrt{\frac{1}{I(I-1)} \sum_{i=1}^{I} (\tilde{\theta}_k - \tilde{\theta})^2}} \approx N(0,1), \tag{3.78}$$

where $\tilde{\theta} = \sum_{i=1}^{I} \tilde{\theta}_k / I$. Thus with probability approximately $1 - \alpha$,

$$\theta \in \tilde{\theta} \pm z^{\alpha/2} \sqrt{\frac{1}{I(I-1)} \sum_{i=1}^{I} (\tilde{\theta}_k - \tilde{\theta})^2}, \tag{3.79}$$

where $z^{\alpha/2}$ is the upper $100(\alpha/2)$ percentile of a unit normal distribution.

For $\theta = (\sigma_e^2 + n\sigma_a^2)/\sigma_e^2 = 1 + n(\sigma_a^2/\sigma_e^2)$ the same procedure can be applied with $\hat{\theta} = MS(A)/MS(E)$. Jackknifing tends to reduce the bias in $\hat{\theta}$ as well as provide robust confidence intervals. In variance ratio problems use of the log transformation – that is, $\hat{\theta} = \log(MS(A)/MS(E))$; $\theta = \log((\sigma_e^2 + n\sigma_a^2)/\sigma_e^2)$ – is likely to improve the normal approximation (3.79). Any confidence interval for θ can be converted to a confidence interval for σ_a^2/σ_e^2 by subtracting 1 from the endpoints and dividing by n.

Arvesen (1969) and Arvesen and Layard (1975) have considered the modifications necessary for handling jackknifing in unbalanced designs.

Jackknifing is not likely to work well on the nonsmooth alternative estimators (3.54)–(3.56). Unless an estimator admits a power series expansion in certain basic variables, the jackknife technique is likely to go awry (see Miller, 1964, 1974a). The jackknife should do well on smooth formal Bayes estimators such as those of Portnoy (1971; see Arvesen, 1969, p. 2092).

3.7. Unequal Variances.

Under $H_0 : \sigma_a^2 = 0$, the effect on the robustness of the F test if σ_e^2 varies from population to population is the same as for the fixed effects model. The reader is referred back to Section 3.3.1. For a balanced design (i.e., $n_i \equiv n$) the effects are minimal, but the distortion can be serious for unbalanced experiments.

The effects on various point and interval estimates for σ_a^2 when $\sigma_{e,i}^2 \not\equiv \sigma_e^2$ are unknown. I have not seen any work concerning the effects on the empirical Bayes estimators of $\{\mu + a_i\}$. The effects on the estimators $\bar{y}_{..}$ and $\bar{y}_{..}^*$ of μ are calculable; see P. S. R. S. Rao et al. (1981) for numerical results.

Detection would be the same as for the fixed effects model (see Sections 3.3.2 and 3.3.3). Plotting s_i versus $\bar{y}_{i\cdot}$ is the best hope of detecting systematic change.

Since estimation of σ_e^2 and/or σ_a^2 is often the primary problem, use of nonparametric techniques is obviated as a corrective device. Transformation may even perturb the problem too much to be useful. P. S. R. S. Rao et al. (1981) study estimators modified for unequal $\sigma_{e,i}^2$.

3.8. Dependence.

As opposed to the models previously considered in this book, dependence between observations is already present. Since

$$\text{Cov}(y_{ij}, y_{ik}) = E[(a_i + e_{ij})(a_i + e_{ik})] = \sigma_a^2 \tag{3.80}$$

for $j \neq k$, the correlation between two observations from the same population is

$$\frac{\sigma_a^2}{\sigma_a^2 + \sigma_e^2}. \tag{3.81}$$

This within population correlation coefficient is called the *intraclass correlation coefficient*, and it is a parameter that has been studied classically in statistics (see Kendall and Stuart, 1961, pp. 302–304). Observations in different populations are, of course, independent under the model.

Blocking because of the presence of a nuisance effect is easily handled through a higher-way classification model, but any other kind of dependence outside the model spells big trouble. Serial correlation between the e_{ij}, or between the a_i, can have a substantial effect. Unfortunately, little or nothing has been written on this or on what to do about it.

Exercises.

1. For the one-way classification with fixed effects, show that

$$E[MS(A)] = \sigma^2 + \frac{\sum_{i=1}^{I} n_i \alpha_i^2}{I-1}.$$

2. For the balanced one-way classification with fixed effects, show that $SS(A) \sim \sigma^2\, \chi^2_{I-1}(\delta^2)$, where $\delta^2 = n \sum_{i=1}^{I} \alpha_i^2/\sigma^2$.

3. Let $y_1, \cdots, y_n$ be independently, identically, continuously distributed, and let R_i be the rank of y_i in the sample. Show that

   (a) $E(R_i) = (n+1)/2$,

   (b) $\mathrm{Var}(R_i) = (n+1)(n-1)/12$,

   (c) $\mathrm{Cov}(R_i, R_{i'}) = (n+1)/12$, $i \neq i'$.

   Hint: $P\{R_i = k\} = 1/n$, $k = 1, \cdots, n$.

4. Use the results of Exercise 3 to establish that for a one-way classification rank analysis

   (a) $E(\bar{R}_{i\cdot}) = (N+1)/2$,

   (b) $\mathrm{Var}(\bar{R}_{i\cdot}) = [N(N+1)/12n_i] - (N+1)/12$,

   (c) $\mathrm{Cov}(\bar{R}_{i\cdot}), \bar{R}_{i'\cdot}) = (N+1)/12$,

   where $N = \sum_{i=1}^{I} n_i$ and $(\bar{R}_{1\cdot}, \cdots, \bar{R}_{I\cdot})$ is the average rank vector [see the discussion preceding (3.26)].

5. Use the results of Exercise 4 to establish that for the linear rank statistic $L = \sum_{i=1}^{I} i\, \bar{R}_{i\cdot}$ [see the discussion preeding (3.37)]

   (a) $E(L) = \frac{(N+1)I(I+1)}{4}$,

(b) $\text{Var}(L) = \frac{N(N+1)}{12}\left(\sum_{i=1}^{I} \frac{i^2}{n_i}\right) - \frac{(N+1)I^2(I+1)^2}{48}$.

6. For the one-way classification with random effects, show that

$$E[MS(A)] = \sigma_e^2 + \frac{1}{N(I-1)}\left(N^2 - \sum_{i=1}^{I} n_i^2\right)\sigma_a^2.$$

7. For the balanced one-way classification with random effects, show that $SS(A) \sim (\sigma_e^2 + n\sigma_a^2)\chi_{I-1}^2$.

8. Prove that $S/(\nu+2)$ minimizes the mean squared error among the class of estimators cS for σ^2, where c is a constant and $S \sim \sigma^2\,\chi_\nu^2$.

9. A clinical method for evaluating trunk flexor muscle strength in children was needed to assist physical therapists in accurately assessing strength in pediatric patients. In this Stanford study trunk flexor muscle strength was measured in 75 girls 3 to 7 years of age.* Muscle strength was graded on a scale of 0 to 5 using modified manual muscle testing methods. These methods attempted to minimize the amount of hip flexor muscle activity during trunk flexion while allowing more isolated action of the abdominal trunk flexors.

   The means and standard deviations ($\bar{y} \pm s$) for the girls grouped by years of age ($n = 15$ in each group) are summarized in the table.

Age	3	4	5	6	7
Muscle Grade	3.3± 0.9	3.7± 1.1	4.1± 1.1	4.4± 0.9	4.8 ± 0.5

   (a) Run an ANOVA test of the null hypothesis of no age effects.

   (b) Use Tukey studentized range intervals to decide which age groups differ.

* Baldauf, K. L., Swenson, D. K., Medeiros, J. M., and Radtka, S. A. (1984). Clinical assessment of trunk flexor muscle strength in healthy girls 3 to 7 years of age. *Physical Therapy*, **64**, 1203–1208.

(c) Apply a linear-2 contrast test for monotone alternatives to test for muscle grade increasing with age.

10. Plasma bradykininogen levels were measured in normal subjects, in patients with active Hodgkin's disease, and in patients with inactive Hodgkin's disease. The globulin bradykininogen is the precursor substance for bradykinin, which is thought to be a chemical mediator of inflammation. The data (in micrograms of bradykininogen per milliliter of plasma) are displayed in the table. The medical investigators wanted to know if the three groups differed in their bradykininogen levels.* Carry out the statistical analysis you consider to be most appropriate, and state your conclusions on this question.

* Eilam, N., Johnson, P. K., Johnson, N. L., and Creger, W. P. (1968). Bradykininogen levels in Hodgkin's disease. *Cancer*, **22**, 631–634.

Normal Controls	Active Hodgkin's Disease	Inactive Hodgkin's Disease
5.37	3.96	5.37
5.80	3.04	10.60
4.70	5.28	5.02
5.70	3.40	14.30
3.40	4.10	9.90
8.60	3.61	4.27
7.48	6.16	5.75
5.77	3.22	5.03
7.15	7.48	5.74
6.49	3.87	7.85
4.09	4.27	6.82
5.94	4.05	7.90
6.38	2.40	8.36
9.24	5.81	5.72
5.66	4.29	6.00
4.53	2.77	4.75
6.51	4.40	5.83
7.00		7.30
6.20		7.52
7.04		5.32
4.82		6.05
6.73		5.68
5.26		7.57
		5.68
		8.91
		5.39
		4.40
		7.13

11. In an experiment on the effects of oxygen toxicity in newborn mice, littermates were separated at birth into chambers containing air or nearly 100% oxygen. Pairs of nursing mothers were switched between the chambers every 12 hours to avoid oxygen intoxication of the mothers. This experiment was repeated 4 times with the newborn mice in the chambers for 24 hours. The amounts of tritiated thymidine incorporated into the pulmonary DNA (dpm/μg DNA) in the air and O_2-exposed mice are displayed in the table. Additional experiments were run for 36, 48, and 72 hours.*

    Estimate the variance component in the differences due to experiments by the method of moments from the ANOVA Table 1 [i.e., (50) with $I = 4$, $n = 4$]. Assume no nursing pair effect.

---

* Northway, W. H., Jr., Petriceks, R., and Shahinian, L. (1972). Quantitative aspects of oxygen toxicity in the newborn: Inhibition of lung DNA synthesis in the mouse. *Pediatrics*, **50**, 67–72.

Experiment	Nursing Pair	Litter Mother	Air	O_2	Difference
E_1	M_1M_3	M_1	11.2	23.9	-12.7
		M_3	26.1	7.5	18.6
	M_2M_4	M_2	14.2	16.6	-2.4
		M_4	7.3	14.3	-7.0
E_2	M_1M_3	M_1	17.4	19.3	-1.9
		M_3	16.8	14.9	1.9
	M_2M_4	M_2	15.6	1.6	14.0
		M_4	12.6	4.6	8.0
E_3	M_1M_3	M_1	12.6	4.6	8.0
		M_3	20.4	11.2	9.2
	M_2M_4	M_2	5.6	8.8	-3.2
		M_4	19.2	16.4	2.8
E_4	M_1M_3	M_1	11.2	7.8	3.4
		M_3	13.5	9.8	3.7
	M_2M_4	M_2	12.6	13.3	-0.7
		M_4	7.4	5.4	2.0

Chapter 4

TWO-WAY CLASSIFICATION

With a two-way classification there are two distinct factors affecting the observed responses. Each factor is investigated at a variety of different levels in an experiment, and the combinations of the two factors at different levels form a cross-classification.

The simplest linear model for an observation y_{ij} taken at level i of Factor A and level j of Factor B is

$$y_{ij} = \mu + \alpha_i + \beta_j + e_{ij}, \tag{4.1}$$

where μ is the overall mean, e_{ij} is the unexplained variation, and α_i and β_j are the effects for Factors A and B, respectively. The more general model

$$y_{ij} = \mu + \alpha_i + \beta_j + \alpha\beta_{ij} + e_{ij} \tag{4.2}$$

allows for an interactive effect $\alpha\beta_{ij}$ between the Factors A and B at the levels combination (i, j). Sometimes more than one observation is taken at the (i, j) combination of levels so a third indexing subscript k is added to y and e (i.e., y_{ijk} and e_{ijk}).

The assumptions that should be imposed on $\{\alpha_i\}$, $\{\beta_j\}$, and $\{\alpha\beta_{ij}\}$ are dictated by the types of factors involved in the experiment. As in the one-way classification, it is necessary to distinguish between *fixed* and *random* effects. Different treatments, types of disease, age groupings, sex, etc., are typically considered to be fixed effects, and the statistical inference extends only to those included in the experiment. On the other hand, patients, days, batches, etc., are usually considered to be merely representatives from a larger population; thus they are handled as random effects.

In a two-way classification each factor can be either fixed or random. If both factors are fixed, the model is called a *fixed effects model*. When both are random, it is called a *random effects model*, and when there is one of each, it is a *mixed effects model*. In Churchill Eisenhart's (1947) terminology, these are referred to as Models I, II, and III, respectively. The sections of this chapter discuss the fixed, mixed, and random effects models.

FIXED EFFECTS

4.1. Normal Theory.

When Factor A has I levels (i.e., $i = 1, \cdots, I$) and Factor B has J levels (i.e., $j = 1, \cdots, J$), the cross-array of (i, j) combinations has IJ cells. Let μ_{ij} be the mean for the cell (i, j). Any arbitrary set of IJ means $\{\mu_{ij}\}$ can be expressed in the form

$$\mu_{ij} = \mu + \alpha_i + \beta_j + \alpha\beta_{ij}, \tag{4.3}$$

where the constraints

$$\begin{gathered} \sum_{i=1}^{I} \alpha_i = 0, \quad \sum_{j=1}^{J} \beta_j = 0, \\ \sum_{i=1}^{I} \alpha\beta_{ij} = 0 \text{ for all } j, \quad \sum_{j=1}^{J} \alpha\beta_{ij} = 0 \text{ for all } i, \end{gathered} \tag{4.4}$$

are imposed on the α, β, and $\alpha\beta$ parameters. These parameters, subject to the constraints (4.4) are defined in terms of the μ_{ij} as follows:

$$
\begin{aligned}
\mu &= \frac{1}{IJ}\sum_{i=1}^{I}\sum_{j=1}^{J}\mu_{ij}, \\
\alpha_i &= \frac{1}{J}\sum_{j=1}^{J}\mu_{ij} - \mu, \\
\beta_j &= \frac{1}{I}\sum_{i=1}^{I}\mu_{ij} - \mu, \\
\alpha\beta_{ij} &= \mu_{ij} - \alpha_i - \beta_j - \mu.
\end{aligned} \tag{4.5}
$$

When the model (4.1) is selected for the analysis, a strong restriction is imposed on the structure of the μ_{ij}; namely, the effects of the two factors must be strictly *additive*. Whether this assumption is warranted in an experiment needs to be carefully considered. Models intermediate between the strictly additive and completely arbitrary models can and will be studied.

4.1.1. Analysis of Variance (ANOVA)

To start, consider the balanced full model

$$
y_{ijk} = \mu + \alpha_i + \beta_j + \alpha\beta_{ij} + e_{ijk} \tag{4.6}
$$

with n replicate observations per cell (i.e., $k = 1, \cdots, n$). Because the parameter sets $\{\alpha_i\}$, $\{\beta_j\}$, $\{\alpha\beta_{ij}\}$ are completely orthogonal in this balanced design, the likelihood ratio tests of the null hypotheses

$$
H_0 : \alpha\beta_{ij} \equiv 0, \quad H_0 : \beta_j \equiv 0, \text{ and } H_0 : \alpha_i \equiv 0 \tag{4.7}
$$

lead to the analysis of variance displayed in Table 4.1.

Customarily, $SS(A)$ is computed in the form $\left(nJ\sum_{i=1}^{I}\bar{y}_{i\cdot\cdot}^2\right) - nIJ\bar{y}_{\cdot\cdot\cdot}^2$, $SS(B)$ in the form $\left(nI\sum_{j=1}^{J}\bar{y}_{\cdot j\cdot}^2\right) - nIJ\bar{y}_{\cdot\cdot\cdot}^2$, $SS(E)$ in the form $\left(\sum_{i=1}^{I}\sum_{j=1}^{J}\sum_{k=1}^{n}y_{ijk}^2\right) - \left(n\sum_{i=1}^{I}\sum_{j=1}^{J}\bar{y}_{ij\cdot}^2\right)$, and $SS(AB)$ by subtraction.

Table 4.1. ANOVA Table for a balanced two-way classification. [a]

VDT	df	SS
Mean (M)	1	$nIJ\,\bar{y}_{\cdot\cdot\cdot}^2$
Factor (A)	$I-1$	$nJ\sum_{i=1}^{I}(\bar{y}_{i\cdot\cdot}-\bar{y}_{\cdot\cdot\cdot})^2$
Factor (B)	$J-1$	$nI\sum_{j=1}^{J}(\bar{y}_{\cdot j\cdot}-\bar{y}_{\cdot\cdot\cdot})^2$
Interactions (AB)	$(I-1)(J-1)$	$n\sum_{i=1}^{I}\sum_{j=1}^{J}(\bar{y}_{ij\cdot}-\bar{y}_{i\cdot\cdot}-\bar{y}_{\cdot j\cdot}+\bar{y}_{\cdot\cdot\cdot})^2$
Error (E)	$IJ(n-1)$	$\sum_{i=1}^{I}\sum_{j=1}^{J}\sum_{k=1}^{n}(y_{ijk}-\bar{y}_{ij\cdot})^2$
Total	IJn	$\sum_{i=1}^{I}\sum_{j=1}^{J}\sum_{k=1}^{n}y_{ijk}^2$

The mean squares column (i.e., $MS = SS/df$) is usually also printed out in packaged computer programs along with corresponding F ratios.

The distribution theory for the sums of squares in Table 4.1 is

---

[a] $\bar{y}_{ij\cdot} = \frac{1}{n}\sum_{k=1}^{n}y_{ijk}$,
$\bar{y}_{i\cdot\cdot} = \frac{1}{nJ}\sum_{j=1}^{J}\sum_{k=1}^{n}y_{ijk}$,
$\bar{y}_{\cdot j\cdot} = \frac{1}{nI}\sum_{i=1}^{I}\sum_{k=1}^{n}y_{ijk}$,
$\bar{y}_{\cdot\cdot\cdot} = \frac{1}{nIJ}\sum_{i=1}^{I}\sum_{j=1}^{J}\sum_{k=1}^{n}y_{ijk}$.

very similar to the one-way classification:

$$
\begin{aligned}
SS(M) &\sim \sigma^2 \chi_1^2 \left(\frac{nIJ\,\mu^2}{\sigma^2} \right), \\
SS(A) &\sim \sigma^2 \chi_{I-1}^2 \left(\frac{nJ \sum_{i=1}^I \alpha_i^2}{\sigma^2} \right), \\
SS(B) &\sim \sigma^2 \chi_{J-1}^2 \left(\frac{nI \sum_{j=1}^J \beta_j^2}{\sigma^2} \right), \\
SS(AB) &\sim \sigma^2 \chi_{(I-1)(J-1)}^2 \left(\frac{n \sum_{i=1}^I \sum_{j=1}^J \alpha\beta_{ij}^2}{\sigma^2} \right), \\
SS(E) &\sim \sigma^2 \chi_{IJ(n-1)}^2,
\end{aligned}
\tag{4.8}
$$

and the five sums of squares are independent. The expected mean squares are

$$
\begin{aligned}
E(MS(M)) &= \sigma^2 + \frac{nIJ\,\mu^2}{\sigma^2}, \\
E(MS(A)) &= \sigma^2 + \frac{nJ \sum_{i=1}^I \alpha_i^2}{(I-1)\sigma^2}, \\
E(MS(B)) &= \sigma^2 + \frac{nI \sum_{j=1}^J \beta_j^2}{(J-1)\sigma^2}, \\
E(MS(AB)) &= \sigma^2 + \frac{n \sum_{i=1}^I \sum_{j=1}^J \alpha\beta_{ij}^2}{(I-1)(J-1)\sigma^2}, \\
E(MS(E)) &= \sigma^2.
\end{aligned}
\tag{4.9}
$$

The appropriate F statistics for testing the null hypotheses in (4.7) by the likelihood ratio method are, respectively,

$$
F = \frac{MS(AB)}{MS(E)}, \quad F = \frac{MS(B)}{MS(E)}, \text{ and } F = \frac{MS(A)}{MS(E)}. \tag{4.10}
$$

In their numerators the F statistics in (4.10) have $(I-1)(J-1)$, $(J-1)$, and $(I-1)$ degrees of freedom, respectively. Their common denominator has $IJ(n-1)$ df. Each F statistic in (4.10) has a central F distribution under the associated null hypothesis and a noncentral F distribution under the alternative.* The tests reject for large values of F so the upper tail of the F distribution gives the P value. This P value is multisided because the alternatives are general. There is no analog to the one-sided P values of Chapters 1 and 2.

If the estimated interaction effects (i.e., $\widehat{\alpha\beta}_{ij} = \bar{y}_{ij\cdot} - \bar{y}_{i\cdot\cdot} - \bar{y}_{\cdot j\cdot} + \bar{y}_{\cdots}$) are statistically significant, the interpretation of the estimated main effects (i.e., $\hat{\alpha}_i = \bar{y}_{i\cdot\cdot} - \bar{y}_{\cdots}$ and $\hat{\beta}_j = \bar{y}_{\cdot j\cdot} - \bar{y}_{\cdots}$) becomes less straightforward than if the interactions are insignificant. The presence of interactions means that, for example, a treatment effect (i.e., level of Factor A) has to be evaluted in terms of the conditions or types of patients (i.e., level of Factor B) to which it is to be applied. The interactions could be so large as to switch the treatment of choice depending upon the conditions or patients. Mere statistical significance of the estimated main effects is not enough to substantiate the superiority of one or more treatments. On the other hand, the estimated interactions can be statistically significant but insufficient in size in comparison to the estimated main effects to cloud the issue. To ascertain their impact, one has to examine the sets of estimates $\{\hat{\alpha}_i\}$, $\{\hat{\beta}_j\}$, and $\{\widehat{\alpha\beta}_{ij}\}$ as well as the sums of squares.

Consider next the case of an unbalanced design where the number of replicates n_{ij} in cell (i, j) varies with the cell. Assume $n_{ij} \geq 1$ for all cells.

If the two-way classification is badly unbalanced with the cell sample sizes differing by orders of magnitude (e.g., 10 or more observations in some cells and only 1 or 2 in others), the prudent analysis is to resort to multiple regression on a large computer. The $\boldsymbol{X}$ matrix

---

* A noncentral $F_{\nu_1,\nu_2}(\delta^2)$ variable (or distribution) is distributed as $(\chi^2_{\nu_1}(\delta^2)/\nu_1)/(\chi^2_{\nu_2}/\nu_2)$, where the noncentral $\chi^2_{\nu_1}(\delta^2)$ variable and the central $\chi^2_{\nu_2}$ variable are independent.

in the regression model $\boldsymbol{Y} = \boldsymbol{X\beta} + \boldsymbol{e}$ for the two-way classification should be constructed of 1's, 0's, and -1's to insert or leave out the appropriate parameters for each cell and to incorporate the constraints (4.4) by expressing some parameters as negative sums of the others. For greater detail on this approach the reader can read Draper and Smith (1981, Chapter 9).

Unfortunately, even with this subterfuge, the analysis is murkier than in the balanced case. The parameter sets are no longer orthogonal, so the sequence in which the hypotheses in (4.7) are tested makes a difference. For example, one has to decide whether one is going to test the $SS(A)$ adjusted for $\alpha\beta$ and β (or perhaps just adjusted for β if the $\alpha\beta$ are insignificant) or the $SS(A)$ unadjusted against $SS(E)$. The size and significance of the other factors affect the choice of test for a factor. No single partition of sums of squares is possible. The regression program has to be run repeatedly with sets of parameters inserted or deleted to obtain the appropriate sums of squares for differencing. Some packaged computer programs will do this for you either automatically or with the proper commands.

One hopes to avoid this predicament and be in a position where the following approximate analysis suffices. Compute each cell mean $\bar{y}_{ij\cdot}$ from all the observations in the cell, and, similarly, compute the error sum of squares from all the observations in the cells of the two-way classification:

$$\begin{aligned} SS(E) &= \sum_{i=1}^{I}\sum_{j=1}^{J}\sum_{k=1}^{n_{ij}}(y_{ijk} - \bar{y}_{ij\cdot})^2, \\ &= \sum_{i=1}^{I}\sum_{j=1}^{J}\sum_{k=1}^{n_{ij}} y_{ijk}^2 - \sum_{i=1}^{I}\sum_{j=1}^{J} n_{ij}\bar{y}_{ij\cdot}^2 . \end{aligned} \tag{4.11}$$

However, in computing the other entries in the ANOVA table, the $\bar{y}_{ij\cdot}$ are treated as though they were all averages of n^* observations where the n^* is the harmonic mean of the $\{n_{ij}\}$:

$$n^* = \left(\frac{1}{IJ} \sum_{i=1}^{I} \sum_{j=1}^{J} n_{ij}^{-1} \right)^{-1} . \tag{4.12}$$

This leads to the approximate analysis of variance displayed in Table 4.2.

Table 4.2. Approximate ANOVA for an unbalanced two-way classification [a]

VDT	df	SS
Mean (M)	1	$n^* IJ (\bar{y}^*_{\cdots})^2$
Factor (A)	$I-1$	$n^* J \sum_{i=1}^{I} (\bar{y}^*_{i\cdot\cdot} - \bar{y}^*_{\cdots})^2$
Factor (B)	$J-1$	$n^* I \sum_{j=1}^{J} (\bar{y}^*_{\cdot j\cdot} - \bar{y}^*_{\cdots})^2$
Interactions (AB)	$(I-1)(J-1)$	$n^* \sum_{i=1}^{I} \sum_{j=1}^{J} (\bar{y}_{ij\cdot} - \bar{y}^*_{i\cdot\cdot} - \bar{y}^*_{\cdot j\cdot} + \bar{y}^*_{\cdots})^2$
Error (E)	$N - IJ$	$\sum_{i=1}^{I} \sum_{j=1}^{J} \sum_{k=1}^{n_{ij}} (y_{ijk} - \bar{y}_{ij\cdot})^2$

The sums of squares in Table 4.2 do not add exactly to the total sum of squares, but the discrepancy should not be too great. Only $SS(E)/\sigma^2$ has precisely a χ^2 distribution. All the other sums of squares (divided by σ^2) have approximate noncentral (or central under H_0) χ^2 distributions. The sum of squares $SS(E)$ is indepen-

[a] $\bar{y}_{ij\cdot} = \frac{1}{n_{ij}} \sum_{k=1}^{n_{ij}} y_{ijk}$,
$\bar{y}^*_{i\cdot\cdot} = \frac{1}{J} \sum_{j=1}^{J} \bar{y}_{ij\cdot}$,
$\bar{y}^*_{\cdot j\cdot} = \frac{1}{I} \sum_{i=1}^{I} \bar{y}_{ij\cdot}$,
$\bar{y}^*_{\cdots} = \frac{1}{IJ} \sum_{i=1}^{I} \sum_{j=1}^{J} \bar{y}_{ij\cdot}$,
$N = \sum_{i=1}^{I} \sum_{j=1}^{J} n_{ij}$.

dent of the rest, but the others lose their interindependence. The approximate F tests of (4.7) employ the usual ratios (4.10).

Although the preceding analysis is only approximate, it is easy to carry out and to interpret. Rankin (1974) has shown that it does not give misleading results provided the ratios of sample sizes do not exceed 3. He also studied a modified analysis in which the numerator degrees of freedom are adjusted for the irregularities in sample sizes.

When there is just a single observation per cell (i.e., $n_{ij} \equiv n = 1$), the analysis of variance in Table 4.1 reduces to that in Table 4.3. Notice that the row for "Error" has vanished from this table since there are no replicate observations for measuring error. This leaves the statistician in a pickle because there is no denominator for the F statistics in (4.10).

The statistician has two choices.

The first is to close his or her eyes, cross his or her fingers, and use $SS(AB)$ as an error sum of squares. This leads to

$$F = \frac{MS(B)}{MS(AB)} \quad \text{and} \quad F = \frac{MS(A)}{MS(AB)} \tag{4.13}$$

being used as the test statistics for the last two null hypotheses in (4.7). If there are no interactions, the ratios in (4.13) have (noncentral) F distributions with $J - 1$ and $I - 1$ df in their numerators, respectively, and $(I - 1)(J - 1)$ df in their denominators.

All this is fine provided there are no interactions. In some experiments the assumption $\alpha\beta_{ij} \equiv 0$ may be justified because of the nature of the factors. A synergistic reaction between them would not be conceivable. However, if interactions are indeed present, they inflate the sum of squares in the denominator and unduly dampen the significance of the numerator sum of squares. Of course, if the ratio is significantly large as judged by the central F distribution, the issue of whether the main effects are really even more significant is academic.

Table 4.3. ANOVA table for a balanced ($n = 1$) two-way classification [a]

VDT	df	SS
Mean (M)	1	$IJ\,\bar{y}_{\cdot\cdot}^2$
Factor (A)	$I-1$	$J\sum_{i=1}^{I}(\bar{y}_{i\cdot} - \bar{y}_{\cdot\cdot})^2$
Factor (B)	$J-1$	$I\sum_{j=1}^{J}(\bar{y}_{\cdot j} - \bar{y}_{\cdot\cdot})^2$
Interactions (AB)	$(I-1)(J-1)$	$\sum_{i=1}^{I}\sum_{j=1}^{J}(y_{ij} - \bar{y}_{i\cdot} - \bar{y}_{\cdot j} + \bar{y}_{\cdot\cdot})^2$
Total	IJ	$\sum_{i=1}^{I}\sum_{j=1}^{J} y_{ij}^2$

In other instances it might be argued that the main effects are only of interest if they are substantially larger than the interactions. The F ratios (4.13) reflect the relative sizes of the main effects and interactions, but computing P values from an F distribution under such circumstances is a fantasy. When interactions are present, $SS(AB)$ has a noncentral χ^2 distribution, and the ratios in (4.13) have doubly noncentral F distributions.*

The alternative choice available to the statistician is to try to split $SS(AB)$ into two components of which one soaks up most of the interactive effects and the other is mainly pure error. Tukey

[a] $\bar{y}_{i\cdot} = \frac{1}{J}\sum_{j=1}^{J} y_{ij}$,
$\bar{y}_{\cdot j} = \frac{1}{I}\sum_{i=1}^{I} y_{ij}$,
$\bar{y}_{\cdot\cdot} = \frac{1}{IJ}\sum_{i=1}^{I}\sum_{j=1}^{J} y_{ij}$.

* A doubly noncentral $F_{\nu_1,\nu_2}(\delta_1^2, \delta_2^2)$ variable (or distribution) is distributed as $[\chi_{\nu_1}^2(\delta_1^2)/\nu_1]/[\chi_{\nu_2}^2(\delta_2^2)/\nu_2]$ where the noncentral variables $\chi_{\nu_1}^2(\delta_1^2)$ and $\chi_{\nu_2}^2(\delta_2^2)$ are independent.

(1949) proposed separating from $SS(AB)$ *one degree of freedom for nonadditivity* which would engulf most of the interactive effects in a quadratic model. Specifically, if the response surface is postulated to be quadratic in the main effects, i.e.,

$$\mu_{ij} \propto (\mu + \alpha_i + \beta_j)^2, \tag{4.14}$$

then expansion and rearrangement of terms gives

$$\mu_{ij} \propto \mu^2 + (2\mu\alpha_i + \alpha_i^2) + (2\mu\beta_j + \beta_j^2) + 2\alpha_i\beta_j, \tag{4.15}$$

so the interaction $\alpha_i\beta_j$ is multiplicative in nature. A single square term that is sensitive to detecting interactions of this form is

$$SS_1 = \frac{\left[\sum_{i=1}^{I}\sum_{j=1}^{J}\hat{\alpha}_i\hat{\beta}_j y_{ij}\right]^2}{\sum_{i=1}^{I}\hat{\alpha}_i^2\sum_{j=1}^{J}\hat{\beta}_j^2}. \tag{4.16}$$

Scheffé (1959, Section 4.8) provides a more rigorous derivation of the statistic (4.16) and shows that when there are no interactions, SS_1 and $SS(E) - SS_1$ are statistically independent and have χ^2 distributions with 1 and $(I-1)(J-1)-1$ df, respectively. Thus one can test for the presence of interactions by comparing the ratio

$$\frac{(I-1)(J-1)-1}{1} \cdot \frac{SS_1}{SS(E) - SS_1} \tag{4.17}$$

with the critical values for an F distribution with 1 and $(I-1)(J-1)-1$ df. If interactions are present and generally multiplicative in nature, then SS_1 should soak up most of them and leave $SS(E)-SS_1$ relatively uncontaminated, so it should be possible to use the latter sum of squares for legitimately testing the main effects.

Tukey (1955) and Abraham (1960) extended this idea to Latin squares. In Problem 4.19 Scheffé (1959) indicated how to generalize this method for testing other forms of interactions in the general linear model. Later Milliken and Graybill (1970) elaborated on this generalization.

Occasionally a single observation is missing because, for example, a slide has been dropped or an animal has been lost for reasons unrelated to the experiment. When the experiment is otherwise balanced with $n > 1$, I would run the approximate analysis in Table 4.2 with $n^* = n$. In other words, consider each cell mean as being based on the full n observations and compute $SS(E)$ from the observations available in each cell. However, if $n = 1$, this cannot be done because $\bar{y}_{ij\cdot}$ cannot be computed for the cells with missing data.

For a single observation per cell experiment with a lot of missing data, there is nothing to be done other than to resort to running the data through a multiple regression program. However, with a single missing value in cell (k, ℓ), one can substitute

$$\hat{y}_{k\ell} = \frac{IR_k + JC_\ell - T}{(I-1)(J-1)} \tag{4.18}$$

for the missing observation, where R_k is the sum of the nonmissing observations in row k, C_ℓ is the sum of the nonmissing observations in column ℓ, and T is the total sum of all the nonmissing observations in the $I \times J$ array. The sums of squares given in Table 4.3 can then be calculated, but the df for interactions (AB) should be reduced by one to $(I-1)(J-1)-1$. Approximate F tests can then be performed by computing the usual ratios.

If one desires more accuracy, it is possible to compute an exact analysis of variance without resorting to multiple regression. For details the reader is referred to Kempthorne (1952, pp. 172–174).

An iterative procedure using (4.18) is available when two or more observations are missing and $n = 1$; see Cochran and Cox (1957, pp. 110–112). When $n > 1$, an approximate analysis (see Table 4.2) is usually satisfactory for several missing values provided all cell means can be estimated.

4.1.2. Multiple Comparisons

The idea and methods of multiple comparisons were introduced in Section 3.1.2. The reader may want to refer back to this section or to a fuller discussion in Miller (1981).

Essentially all the methods introduced in Section 3.1.2 extend to the two-way classification with the only change being in what is used for $\hat{\sigma}^2$. If there is more than one observation in all, or at least some, of the cells, then

$$\hat{\sigma}^2 = \frac{SS(E)}{N - IJ}, \tag{4.19}$$

where $SS(E)$ has $\nu = N - IJ$ df and $N = \sum_{i=1}^{I} \sum_{j=1}^{J} n_{ij}$. With just a single observation for each cell, then

$$\hat{\sigma}^2 = \frac{SS(AB)}{(I-1)(J-1)} \tag{4.20}$$

with $\nu = (I-1)(J-1)$ df. The appropriate subtractions should be made in the numerator and denominator of (4.20) for missing observations and deletion of single degrees of freedom for nonadditivity (see Section 4.1.1).

For a balanced design (i.e., $n_i \equiv n$) the *Tukey intervals* are

$$\alpha_i - \alpha_{i'} \in \bar{y}_{i\cdot\cdot} - \bar{y}_{i'\cdot\cdot} \pm q^{\alpha}_{I,\nu} \frac{\hat{\sigma}}{\sqrt{nJ}}, \tag{4.21}$$

where $q^{\alpha}_{I,\nu}$ is the upper 100α percentile of a studentized range distribution for I numerator variables with ν df in the denominator and $\hat{\sigma}$ is given by (4.19) or (4.20). When a design is slightly unbalanced and the approximate analysis given in Table 4.2 is used, then (4.21) can be applied with n^* replacing n. The coverage probability of (4.21) for all pairs i and i' is exactly $1 - \alpha$ in the balanced case and is approximately the same in the unbalanced case. As the design becomes more unbalanced, the coverage probability deteriorates (see Dunnett, 1980a); extension of the Tukey-Kramer intervals (3.9) to the two-way classification should afford better protection in this case.

For badly unbalanced designs where one has to resort to employing multiple regression, the *Scheffé intervals* provide the simultaneous confidence intervals

$$\alpha_i - \alpha_{i'} \in \hat{\alpha}_i - \hat{\alpha}_{i'} \pm [(I-1)F^{\alpha}_{I-1,\nu}]^{1/2}\, \hat{\sigma}[\boldsymbol{c}^T_{ii'}(\boldsymbol{X}^T\boldsymbol{X})^{-1}\boldsymbol{c}_{ii'}]^{1/2}, \quad (4.22)$$

where $\boldsymbol{c}_{ii'}$ is the vector containing 1, -1, and interspersed zeros that pick out the contrast $\alpha_i - \alpha_{i'}$. The error variances $\hat{\sigma}^2$ is the residual sum of squares divided by the degrees of freedom $N - IJ$ or $N - 1 - (I-1) - (J-1)$ depending on whether interactions are included in the model. *Bonferroni intervals* are obtained by substituting $t^{\alpha/2K}_{\nu}$ with $K = \binom{I}{2}$ for $((I-1)F^{\alpha}_{I-1,\nu})^{1/2}$ in (4.22), and these can be shorter than the Scheffé intervals.

For special tables of percentage points to use in conjunction with these methods, see Section 3.1.2.

More general contrasts can be handled as well. The Scheffé and Bonferroni methods merely substitute the appropriate $\boldsymbol{c}$ into $\boldsymbol{c}^T(\boldsymbol{X}^T\boldsymbol{X})^{-1}\boldsymbol{c}$, and Bonferroni must modify K to include the requisite number of contrasts. The Tukey intervals for balanced, or almost balanced, designs must append the multiplicative factor $\sum_{i=1}^{I}|c_i|/2$ [see (3.14)].

Similar intervals could be constructed for Factor B. The symbols β, j, and J are simply substituted for α, i, and I.

If multiple comparisons are made for both Factors A and B, it is not true that the combined coverage would have probability $1 - \alpha$ (or greater). The critical constants would have to be substantially changed to achieve this. However, it is rare that one is interested in multiple comparisons of both factors, and even rarer (never?) that one wants to be so conservative as to have simultaneous coverage on both sets of comparisons.

4.1.3. Monotone Alternatives

The general theory for monotone means of normal distributions discussed in Section 3.1.3 is available for use in the two-way classification.

With balanced, or nearly balanced, designs, either the likelihood ratio approach or the contrast approach can be applied. The only changes from the one-way classification are that $\bar{y}_{i\cdot\cdot}$ based on nJ observations, or $\bar{y}^*_{i\cdot\cdot}$ based on n^*J observations, is substituted for $\bar{y}_{i\cdot}$, and $\hat{\sigma}^2$ from (4.19) or (4.20) is used for the estimate of the variance with its corresponding degrees of freedom.

For an extremely unbalanced design, the likelihood ratio approach fails, not for any theoretical reason, but just for lack of explicit formulas and tables. However, the Abelson-Tukey approach is still possible. For the regression estimates $\hat{\alpha}_1, \cdots, \hat{\alpha}_I$, which have been produced by the computer, one simply calculates the linear combination $\boldsymbol{c}^T\hat{\boldsymbol{\alpha}} = \sum_{i=1}^I c_i\hat{\alpha}_i$, where the c_i are given by (3.21), (3.22), or (3.23). The variance of the contrast is then estimated by $\hat{\sigma}^2\, \boldsymbol{c}^T(\boldsymbol{X}^T\boldsymbol{X})^{-1}\boldsymbol{c}$. If in running the regression program one of the α_i, say, α_I, has been deleted to incorporate the constraint $\sum_{i=1}^I \alpha_i = 0$ by setting $\alpha_I = -\sum_{i=1}^{I-1}\alpha_i$, the contrast should be computed as $\sum_{i=1}^{I-1}(c_i - c_I)\hat{\alpha}_i$ with a corresponding adjustment in the estimated variance.

When the design is balanced, it is possible to test for more general types of monotone alternatives than just linear increases (or decreases). One rarely, if ever, goes beyond quadratic effects to cubic and higher order effects so the discussion is limited to just quadratic effects.

The contrasts that measure a quadratic effect and are orthogonal to the linear contrasts and general mean for $I = 3$ to 7 are displayed in (4.23). These come from the values of the second order orthogonal polynomial that have been normalized into integer form. For additional details on orthogonal polynomials and their construction, the reader can study Draper and Smith (1981, Sections 5.6–5.7) or other sources.

$$\begin{array}{llllllll}
I=3: & & & +1 & -2 & +1 & & \\
I=4: & & +1 & -1 & -1 & +1 & & \\
I=5: & +2 & -1 & -2 & -1 & +2 & & \\
I=6: & +5 & -1 & -4 & -4 & -1 & +5 & \\
I=7: & +5 & 0 & -3 & -4 & -3 & 0 & +5
\end{array} \tag{4.23}$$

As with linear constrasts, one simply calculates $\sum_{i=1}^{I} c_i\bar{y}_{i\cdot\cdot}$ and the estimated variance is $\hat{\sigma}^2 \sum_{i=1}^{I} c_i^2/Jn$, where $\hat{\sigma}^2$ is given by (4.19) or (4.20).

Since the linear and quadratic constrasts are orthogonal, it is possible to subdivide $SS(A)$ into linear, quadratic, and remainder sums of squares. Let $(\ell_1, \cdots, \ell_I)$ denote the linear contrast from (3.21) and $(q_1, \cdots, q_I)$ the quadratic constant from (4.23). Then,

$$SS(A) = \frac{nJ\left(\sum_{i=1}^{I} \ell_i\bar{y}_{i\cdot\cdot}\right)^2}{\sum_{i=1}^{I} \ell_i^2} + \frac{nJ\left(\sum_{i=1}^{I} q_i\bar{y}_{i\cdot\cdot}\right)^2}{\sum_{i=1}^{I} q_i^2} + RSS(A), \tag{4.24}$$

where the last term $RSS(A)$ is obtained by subtraction. The three sums of squares on the right hand side in (4.24) are independently distributed, and under H_0 have central χ^2 distributions with 1, 1, and $I-3$ df, respectively. Each can be tested against $MS(E)$. Significance of the first and/or second sum would indicate the presence of a linear and/or quadratic effect, and significance of the third sum would substantiate the existence of other effects.

If the column effects are considered to have monotone alternatives as well as the rows (or instead of the rows), the same analysis can be applied to the columns with $\hat{\beta}_j$ playing the role of $\hat{\alpha}_i$, J for I, etc.

The possibility of interaction between Factor A and Factor B with monotone alternatives can be tested also. The size of the linear contrast in column j is $\sum_{i=1}^{I} \ell_i \bar{y}_{ij\cdot}$, where $(\ell_1, \cdots, \ell_I)$ is given by (3.21). Since the overall linear contrast is $\sum_{i=1}^{I} \ell_i \bar{y}_{i\cdot\cdot}$, the effect of the jth column on the linear contrast is measured by the difference

$$\begin{aligned} \sum_{i=1}^{I} \ell_i \bar{y}_{ij\cdot} - \sum_{i=1}^{I} \ell_i \bar{y}_{i\cdot\cdot} &= \sum_{i=1}^{I} \ell_i (\bar{y}_{ij\cdot} - \bar{y}_{i\cdot\cdot}), \\ &= \sum_{i=1}^{I} \ell_i (\bar{y}_{ij\cdot} - \bar{y}_{i\cdot\cdot} - \bar{y}_{\cdot j\cdot} + \bar{y}_{\cdot\cdot\cdot}), \qquad (4.25) \\ &= \sum_{i=1}^{I} \ell_i \widehat{\alpha\beta}_{ij}. \end{aligned}$$

The sum of squares

$$SS(A_\ell B) = \frac{n \sum_{j=1}^{J} \left[\sum_{i=1}^{I} \ell_i (\bar{y}_{ij\cdot} - \bar{y}_{i\cdot\cdot})\right]^2}{\sum_{i=1}^{I} \ell_i^2} \qquad (4.26)$$

is sensitive to a A linear $\times$ B interaction, and under H_0 it has a χ^2 distribution with $J - 1$ df and is independent of the leftover interaction sum of squares [i.e., $SS(AB) - SS(A_\ell B)$]. It can be tested against $MS(E)$ to determine if interactions of the form A linear $\times$ B are present.

A similar contrast and sum of squares could be constructed for A quadratic $\times B$ interactions. The coefficients $(q_1, \cdots, q_I)$ would be chosen from (4.23). The interaction sum of squares can be further subdivided into $SS(A_\ell B) + SS(A_q B) + [SS(AB) - SS(A_\ell B) - SS(A_q B)]$, where each of the three sums of squares has an independent χ^2 distribution on $J-1$, $J-1$, and $(I-3)(J-1)$ df, respectively, under H_0.

For monotone alternatives in both directions, one can form the A linear $\times$ B linear contrast

$$\begin{aligned}\sum_{j=1}^{J} \ell_j^c \left(\sum_{i=1}^{I} \ell_i^r \bar{y}_{ij\cdot} \right) &= \sum_{i=1}^{I} \ell_i^r \left(\sum_{j=1}^{J} \ell_j^c \bar{y}_{ij\cdot} \right), \\ &= \sum_{i=1}^{I} \sum_{j=1}^{J} \ell_i^r \ell_j^c \widehat{\alpha\beta}_{ij},\end{aligned} \tag{4.27}$$

with corresponding sum of squares

$$SS(A_\ell B_\ell) = \frac{n\left(\sum_{i=1}^{I} \sum_{j=1}^{J} \ell_i^r \ell_j^c \bar{y}_{ij\cdot}\right)^2}{\sum_{i=1}^{I} (\ell_i^r)^2 \sum_{j=1}^{J} (\ell_j^c)^2}, \tag{4.28}$$

where $(\ell_1^r, \cdots, \ell_I^r)$ and $(\ell_1^c, \cdots, \ell_J^c)$ are the appropriate linear contrasts from (3.21). Under H_0, $SS(A_\ell B_\ell)$ has a single df χ^2 distribution, which is independent of the remaining interaction sum of squares. Similarly, $SS(A_q B_\ell)$, $SS(A_\ell B_q)$, $SS(A_q B_q)$, can be separated out from the parent sum of squares $SS(AB)$.

If the design is not fully balanced but is nearly so, the preceding analysis can be carried out with n^* replacing n [see (4.12)] and with $\bar{y}_{ij\cdot}$ being computed from however many observations are present in the (i, j) cell.

If $n_{ij} \equiv n = 1$ and the populations have an a priori ordering, calculation of (4.28) offers an alternative to Tukey's one degree of freedom for nonadditivity.

If there is an actual quantitative variable associated with the rows and/or columns (i.e., Factor A and/or Factor B are quantitative), then the techniques of regression analysis in Chapter 5 are also available and are usually superior.

4.2. Nonnormality.

4.2.1. Effect

The reader is referred back to Section 3.2.1 for the discussion of the effects of nonnormality in the one-way classification because there is little or no difference for the two-way classification. In the balanced, or nearly balanced, two-way classification the tests for row (or column) effects are essentially the same as one-way tests except that the error sum of squares has been corrected to remove the column (or row) effects and, when $n_{ij} > 1$, interactions. Basically, nonnormality has very little effect on the F, studentized range, and linear contrast tests as along as the size of the design (i.e., IJn) is not too small.

The preceding optimistic remarks must be tempered for badly balanced experiments. Heavy-tailed or contaminated distributions may produce unusual observations (outliers) in the thin part of the design and thereby distort the tests and estimates.

Welch (1937) and Pitman (1938) compared the moments of a beta statistic corresponding to an F statistic (4.13) under normal theory and under permutation theory when there is a single observation per cell (i.e., $n_{ij} \equiv 1$) and no interactions (i.e., $\alpha\beta_{ij} \equiv 0$). The agreement was shown to be good, thereby giving credence to the normal theory analysis for general distributions. This work is summarized in Kempthorne (1952, Chapter 8). Related material and discussion appears in Box and Andersen (1955) and Scheffé (1959, Chapter 9 and Section 10.3).

Welch (1937) also studied the permutation moments for a Latin squares analysis.

4.2.2. Detection

It is more difficult to detect nonnormality in the two-way classification than it is in the one-way classification or one and two sample situations. The problem is that each cell in the two-way array represents a different population so there are only a few observations, sometimes just one, for each population. One cannot make probit plots, or perform tests, for each separate population.

The only recourse is to pool residuals from all the cells. When there are multiple observations per cell, one can use the N differences $r_{ijk} = y_{ijk} - \bar{y}_{ij\cdot}$ and make a single probit plot as in Section 1.2.2. Test statistics could be computed, but their ordinary associated significance levels would be fouled up by the dependencies between the r_{ijk} caused by the subtraction of the cell means. However, these dependencies do not cause any substantial difficulty with the probit plot because the empirical distribution function of the residuals is a consistent estimator for the underlying error distribution (see Duan, 1981). The residual distribution function should give an accurate picture of the true error distribution and enable one to decide whether it is sufficiently close to normal.

With just a single observation per cell, it may not be possible to distinguish between nonnormality and interactions. When interactions are assumed not to exist, the residuals $r_{ij} = y_{ij} - \hat{\mu} - \hat{\alpha}_i - \hat{\beta}_j$ can be used in a plot of the residual distribution function as mentioned in the preceding paragraph. However, if here are some unusual values and/or the plotted quantiles do not fall approximately on a straight line, one cannot be sure whether the lack of fit is due to a nonnormal error distribution or the presence of some interaction terms. There is no way to incorporate general interactions and have any residuals left, but one could estimate interactions with special structure (like $\alpha\beta_{ij} \propto \alpha_i\beta_j$ à la Tukey) and calculate the residuals $r_{ij} = y_{ij} - \hat{\mu}_{ij}$, where $\hat{\mu}_{ij}$ is the estimated cell mean including the special interaction term. It is not clear that all this effort would be warranted since the effects of nonnormality are not severe unless the departure is extreme.

4.2.3. Correction

Transformations Transformations are possible, but they do not seem to be as frequently used with the two-way classification as with the one-way or one and two sample problems. The reason is that transforming the data may destroy an additive linear model and create interactions where none existed before.

On the other hand, one may get lucky and reduce both nonnormality and nonadditivity at the same time. For example, with the quadratic model

$$y_{ij} = (\mu + \alpha_i + \beta_j + e_{ij})^2, \tag{4.29}$$

which was mentioned earlier in (4.14) with regard to nonadditivity, a square root transformation will exactly produce an additive linear model and normal errors (provided the e_{ij} are normally distributed).

Nonparametric Techniques It is not often that a nonparametric technique is used in place of an ANOVA analysis for a two-way classification. Nonparametric methods are more work to run and usually do not provide as much information. Also, special structure on the design and model is typically required in order to apply nonparametric methods.

The one technique you will occasionally see is *Friedman's* (1937) *rank test*. It assumes that there is a single observation per cell (i.e., $n_{ij} \equiv 1$). If there are more observations per cell, then the analysis is run on the cell means $\bar{y}_{ij\cdot}$ and any information in the within-cell variation is ignored. Also, the analysis assumes that no interactions are present. With these restrictions, the analysis for the presence of row effects proceeds by replacing each observation y_{ij} in the jth column by its rank R_{ij} in the set of I observations in column j. Then the test statistic is

$$
\begin{aligned}
Q &= \frac{12J}{I(I+1)} \sum_{i=1}^{I} (\bar{R}_{i\cdot} - \bar{R}_{\cdot\cdot})^2, \\
&= \left(\frac{12J}{I(I+1)} \sum_{i=1}^{I} \bar{R}_{i\cdot}^2 \right) - 3J(I+1),
\end{aligned}
\tag{4.30}
$$

where $\bar{R}_{i\cdot} = \sum_{j=1}^{J} R_{ij}/J$ and $\bar{R}_{\cdot\cdot} = \sum_{i=1}^{I} \bar{R}_{i\cdot}/I = (I+1)/2$. The statistic (4.30) is just the usual row sum of squares computed for the ranks with the proper scale factor in the denominator for it to have a limiting χ^2 distribution with $I-1$ df as the number of columns tends to infinity. Tables of the cdf of Q with the small sample sizes $J = 2(1)13$ for $I = 3$, $J = 2(1)8$ for $I = 4$, and $J = 3, 4, 5$ for $I = 5$ appear in Hollander and Wolfe (1973, Table A.15). Tables for $I = 3$, $J = 2(1)15$ and $I = 4$, $J = 2(1)8$ have been given by Owen (1962) and Lehmann (1975).

When ties are present average ranks can be used. If ties occur excessively, the denominator of Q can be modified to account for this. For an exact expression see Hollander and Wolfe (1973, p. 140).

It is possible to make multiple comparisons based on $(\bar{R}_{1\cdot}, \cdots, \bar{R}_{I\cdot})$. For details see Miller (1981, pp. 172–178) or Hollander and Wolfe (1973, pp. 151–154).

For testing against ordered alternatives Page (1963) proposed the statistic $L = \sum_{i=1}^{I} i\bar{R}_{i\cdot}$, where stochastically larger variables are assumed to correspond to increasing i. The mean and variance of L are $I(I+1)^2/4$ and $(I-1)I^2(I+1)^2/144J$, respectively.

When $n_{ij} \equiv n = 1$ and no interactions are present, rank tests other than Friedman's test are also available. Some of these are based on the $\binom{I}{2}$ signed-rank statistics for comparing treatments i and i', $i, i' = 1, \cdots, I$, where the pairing is provided by the columns (see, for example, Hollander and Wolfe, 1973,pp. 167–173). For references to additional tests see Hollander and Wolfe (1973, Chapter 7).

If it is also important to test the null hypothesis of no main effects for the other factor, one of the aforementioned tests can be run with the roles of rows and columns being interchanged. One drawback to rank tests in the two-way classification is that there is no way to test both rows and columns in a single, unified analysis.

For a rank analysis when there are missing observations (i.e., $n_{ij} \leq 1$), see Skillings and Mack (1981).

Analyses that utilize the within-block (column) rankings for all observations when $n_{ij} > 1$ exist but are more complicated. For details the reader is referred to Benard and van Elteren (1953), Noether (1967, Section 7.6), and Brunden and Mohberg (1976). Since I have never used these tests in practice, I cannot comment on their effectiveness.

A different rank approach is the method of *aligned ranks*. This was introduced by Hodges and Lehmann (1962) for the case of just two treatments (i.e., $I = 2$) and was extended to the full two-way classification by Mehra and Sarangi (1967). The idea is to eliminate the block (column) effects by "aligning" the blocks. Usually this is accomplished by subtracting from each observation the median or mean of the column for which it is a member. All $N = \sum_{i=1}^{I} \sum_{j=1}^{J} n_{ij}$ aligned observations are then combined and ranked. An appropriate ANOVA-type statistic for the ranks is selected to measure the differences between the average ranks for the rows (i.e., levels of Factor A). The distribution of this statistic is considered under permutations of the observations within columns. Under some conditions this statistic has an asymptotic χ^2 distribution with $I - 1$ df. As in the preceding analyses, it is necessary to assume that no interactions are present. Lehmann (1975, Section 6.3) gives a clear presentation of the use of aligned ranks in the balanced case $n_{ij} \equiv n = 1$.

Many other nonparametric tests exist beside those already mentioned, but they are rarely used in practice. Included among these are sign tests and permutation tests. For the former see Miller (1981, Section 4.2). With the latter, the significance of the observed F ratio is evaluated with respect to its permutation distribution rather than normal theory. Because the computation required to carry out the analysis is excessive except for the smallest designs, permutation tests are not used in practice for two-way classifications. However, the moment calculations by Welch (1937) and Pitman (1938) under the permutation distribution give credence to the robustness of the F test (see Section 4.2.1).

Robust Estimation Robust methods have not really come to the two-way classification so far. One paper using tests analogous to M-estimators is Schrader and Hettmansperger (1980).

4.3. Unequal Variances.

4.3.1. Effect

The main article on the effects of unequal variances in the two-way classification is Box (1954b), where just the model with no interactions and a single observation per cell is considered. Basically, the effects are not large unless the departure from homoscedasticity is quite extreme. If the variances differ from row to row but are constant over columns, then $J\sum_{i=1}^{I}(\bar{y}_{i\cdot} - \bar{y}_{\cdot\cdot})^2$ is behaving as in a balanced one-way classification (see Section 3.3.1), and for the F test of the null hypothesis $H_0 : \alpha_i \equiv 0$, the actual P value is greater than the nominally stated one (i.e., $P_{\text{actual}} > P_{\text{stated}}$) but not by much. For the test of no column effects $H_0 : \beta_j \equiv 0$, the reverse is true ($P_{\text{actual}} < P_{\text{stated}}$) but again not by much.

4.3.2. Detection

When there is just a single observation per cell (i.e., $n_{ij} \equiv 1$), there is little that can be done to detect unequal variances. If the observed values bounce around more in some rows than others, one might interpret this as unequal variances, particularly if the variability appears to be greater for larger (positive) effects. However, it is impossible to distinguish heteroscedasticity from interactions.

When $n_{ij} > 1$ for each of the cells, then it is possible to compute an error variance $s_{ij\cdot}^2$ in each cell, and the methods of Section 3.3 become available. In particular, plotting $s_{ij\cdot}^2$ vs. $\bar{y}_{ij\cdot}$ will reveal whether there is any change in the variance due to increasing size of the variable. At no time would I consider running a formal test on the equality of the cell variances (see Chapter 7).

4.3.3. Correction

One could apply a transformation to the data to try to stabilize the variances. An appropriate transformation might be suggested by the plot of $s_{ij\cdot}^2$ vs. $\bar{y}_{ij\cdot}$ in designs where $n_{ij} > 1$ (see Section 3.3.3). Howver, there is the danger that transforming the data may destroy an additive model and create interactions. The best of all worlds is to find a transformation that creates normality, stabilizes variances, and eliminates interactions.

The nonparametric tests mentioned in Section 4.2.3, "Nonparametric Techniques," such as Friedman's rank test, should be even less sensitive to heterogenous variances than the F tests, but no research has been done on this to date.

4.4. Dependence.

In designs with multiple observations per cell dependence within cells could be created by the presence of an unaccounted for extra nuisance factor that forms blocks of observations. Observations grouping themselves into clusters is an indication of the existence of such a variable. The remedy for this ailment is relatively straightforward – use a higher-way (e.g., three-way) classification for the analysis.

The problem of serial correlation, created for example by observations being taken in a time sequence, is far more serious in its implications and is far more difficult to detect and correct. The principal paper on the effects of serial correlation in the two-way classification is Box (1954b). See also Andersen et al. (1981).

The case of no interactions and a single observation per cell with a first order serial correlation between rows within columns is studied by Box (1954b). Specifically, suppose that $\text{Cor}(y_{ij}, y_{i+1,j})$ $= \rho_1$ for all i, j, and all other correlations are zero. With this probability structure, the F test of $H_0 : \alpha_i \equiv 0$ is not at all seriously affected. Thus treatment comparisons are not substantially affected by serial correlation between the treatment measurements within a block (i.e., column). On the other hand, the F test of $H_0 : \beta_j \equiv 0$ is drastically affected with $P_{\text{actual}} \gg P_{\text{stated}}$ for $\rho_1 > 0$ and the reverse for $\rho_1 < 0$. Thus serial correlations among the mesurements on each treatment can destroy the validity of treatment comparisons.

If, for example, the time sequence in which the observations are taken is known, one can plot the successive time pairs and see if any association is discernible. The presence of row (or column) effects may, however, obscure the appearance of the time association. Unfortunately, even if detected, there is no known correction for a serial effect.

In repeated measurements designs a replicate within row i for Factor A is a subject who receives all levels (i.e., columns) of Factor B. Use of the same subject for different levels of Factor B produces a correlational structure between columns that is usually assumed to be of a special form. Some of the relevant literature on the analysis of repeated measurements is Geisser and Greenhouse (1958) and Huynh and Feldt (1970, 1980). For a study of general correlations in a two-way design with n replicates per cell see Olkin and Vaeth (1981) and Walters and Rowell (1982).

MIXED EFFECTS

Although it is more customary to discuss the random effects model before the mixed effects model, the order is reversed here because the goals of a mixed effects analysis are so similar to those of fixed effects. Main interest centers on testing the equality of different levels of the fixed effects factor (e.g., Factor A) because these are different treatments, products, etc. The other factor (e.g., Factor B) is a nuisance factor, such as days, subjects, and plots of ground, whose levels are viewed as random because they are representatives of a potentially larger group. Testing and estimation of the levels for the random effects factor are not of prime importance.

4.5. Normal Theory.

The model is

$$y_{ijk} = \mu + \alpha_i + b_j + \alpha b_{ij} + e_{ijk}, \tag{4.31}$$

for $i = 1, \cdots, I$, $j = 1, \cdots, J$, $k = 1, \cdots, n_{ij}$. The fixed effects $\{\alpha_i\}$ are assumed to satisfy the constraint $\sum_{i=1}^{I} \alpha_i = 0$ for identifiability. The distributional assumptions are

$$\begin{aligned} b_j &\text{ independent } N(0, \sigma_b^2), \\ e_{ijk} &\text{ independent } N(0, \sigma_e^2), \\ \{b_j\} &\text{ independent of } \{e_{ijk}\}. \end{aligned} \tag{4.32}$$

What about assumptions on the interactions $\{\alpha b_{ij}\}$? Historically, there was a controversy over the choice of proper conditions. In the original version of his textbook, Mood (1950) assumed that the

ab_{ij} are distributed as $N(0, \sigma_{ab}^2)$, independently of each other and the e_{ijk}. On the other hand, R. L. Anderson and Bancroft (1952) in their textbook assumed normality and independence from the e_{ijk}, but they also imposed the constraint $\sum_{i=1}^{I} ab_{ij} = 0$ for each j. This creates dependence between the ab_{ij} within each j level. The rationale for the constraint had its roots in the fixed effects structure. The consequence of the difference in assumptions is that one is led to different denominator sums of squares in testing for the presence of Factor B main effects (i.e., $H_0 : \sigma_b^2 = 0$).

This issue was more or less resolved by the publication of an article by Cornfield and Tukey (1956). In this article they derived the expected mean squares under sampling from a finite population model. Their results agreed in form with Anderson and Bancroft so imposition of the constraint is usually accepted to be appropriate. Searle (1971, Section 9.7) discusses both models.

Scheffé (1959) has the most general model in which he assumes only that the vectors $(b_j, ab_{1j}, \cdots, ab_{Ij})$, $j = 1, \cdots, J$, are independent multivariate normal random vectors that satisfy $\sum_{i=1}^{I} ab_{ij} = 0$ for each j. This allows $ab_{1j}, \cdots, ab_{Ij}$ to be dependent on b_j. Graybill (1961), on the other hand, assumes that the interactions ab_{ij}, $i = 1, \cdots, I$ are independent of the main block effect b_j. With the assumption that the ab_{ij} are identically distributed, this gives the covariance structure

$$
\begin{aligned}
\operatorname{Var}(ab_{ij}) &= \left(1 - \frac{1}{I}\right)\sigma_{ab}^2, \\
\operatorname{Cov}(ab_{ij}, ab_{i'j}) &= -\frac{1}{I}\sigma_{ab}^2 \text{ for } i \neq i'.
\end{aligned}
\tag{4.33}
$$

There are only inconsequential differences in the distribution theory for the sums of squares between the Scheffé and Graybill models so the Graybill model is adopted here for its simplicity.

For a balanced design (i.e., $n_{ij} \equiv n$) the ANOVA Table 4.1

is retained in the mixed effects *analysis of variance*. The central question is what is the distribution theory for its entries. The answer is

$$
\begin{aligned}
SS(M) &\sim (\sigma_e^2 + nI\sigma_b^2)\chi_1^2\left(\frac{nIJ\mu^2}{\sigma_e^2 + nI\sigma_b^2}\right), \\
SS(A) &\sim (\sigma_e^2 + n\sigma_{\alpha b}^2)\chi_{I-1}^2\left(\frac{nJ\sum_{i=1}^{I}\alpha_i^2}{\sigma_e^2 + n\sigma_{\alpha b}^2}\right), \\
SS(B) &\sim (\sigma_e^2 + nI\sigma_b^2)\chi_{J-1}^2, \\
SS(AB) &\sim (\sigma_e^2 + n\sigma_{\alpha b}^2)\chi_{(I-1)(J-1)}^2, \\
SS(E) &\sim \sigma_e^2\chi_{IJ(n-1)}^2,
\end{aligned}
\tag{4.34}
$$

and the five sums of squares are independent.

The null hypotheses of no interactions and no Factor B (column) effects are now stated in terms of variance components, namely, $H_0 : \sigma_{\alpha b}^2 = 0$ and $H_0 : \sigma_b^2 = 0$, respectively. For these two hypotheses one uses the same F ratios as in the fixed effects case, namely,

$$
F = \frac{MS(AB)}{MS(E)} \quad \text{and} \quad F = \frac{MS(B)}{MS(E)}, \tag{4.35}
$$

respectively. The only difference from the fixed effects case is in the calculation of power. Under the alternative hypotheses

$$
\begin{aligned}
\frac{MS(AB)}{MS(E)} &\sim \left(1 + \frac{n\sigma_{\alpha b}^2}{\sigma_e^2}\right)F_{(I-1)(J-1),IJ(n-1)}, \\
\frac{MS(B)}{MS(E)} &\sim \left(1 + \frac{nI\sigma_b^2}{\sigma_e^2}\right)F_{J-1,IJ(n-1)},
\end{aligned}
\tag{4.36}
$$

where the F distributions are central F distributions with their respective df, whereas for fixed effects these rations would have noncentral F distributions and no multiplicative factors.

For testing the null hypothesis $H_0 : \alpha_i \equiv 0$ of no Factor A (row) effects, the test statistic is different from the fixed effects ratio $MS(A)/MS(E)$. Because of the multiplicative factor $\sigma_e^2 + n\sigma_{\alpha b}^2$ in the distribution of $SS(A)$ [see (4.34)], it is necessary to divide by a sum of squares with the same factor. The distribution of the interacation sum of squares $SS(AB)$ has this factor so the ratio

$$F = \frac{MS(A)}{MS(AB)} \tag{4.37}$$

is the appropriate statistic. Under H_0 the ratio (4.37) has a central F distribution with $I-1$, $(I-1)(J-1)$ df for numerator and denominator, respectively, and under the alternative hypothesis it has a noncentral F distribution with the noncentrality parameter $nJ\sum_{i=1}^{I}\alpha_i^2/(\sigma_e^2 + n\sigma_{\alpha b}^2)$.

If one feels rather sure that no interactions are present (i.e., $\sigma_{\alpha b}^2 = 0$), then it is possible to use the fixed effects ratio $MS(A)/MS(E)$ for testing $H_0 : \alpha_i \equiv 0$. This usually provides more degrees of freedom for the denominator. One could even pool the $SS(AB)$ and $SS(E)$ if degrees of freedom are scarce. To do this I would need to have $MS(AB)$ nearly equal to $MS(E)$ and not merely have nonsignificance for $MS(AB)/MS(E)$.

Under Scheffé's more general model for the block (column) and interaction effects, the distribution theory of $SS(A)$ and $SS(AB)$ is more complicated. Their ratio does not have an F distribution. The only way to obtain an exact test of $H_0 : \alpha_i \equiv 0$ is to convert the problem into one in multivariate analysis, and this leads to Hotelling's T^2 test (see Scheffé, 1959, pp. 270–274). However, Scheffé eschews this procedure and suggests the use of the ratio (4.37) as an approximate test under his model.

The preceding analysis of the mixed effects model has been based on the assumption of a balanced design. What if the n_{ij} are not all equal? For mild imbalance I would recommend using the approximate ANOVA presented in Table 4.2,with n^* given by (4.12), in conjunction with the preceding analysis for a mixed model. If you asked me what to do for badly unbalanced designs with random block and interaction effects present, I would probably shrug my shoulders and say "I don't know" or "Use a fixed effects analysis." Searle (1971, Chapters 10 and 11) struggles with this problem but has no simple solution.

The design with $n_{ij} \equiv n = 1$ causes less of a dilemma for the mixed effects model than it does for the fixed effects model. With mixed effects the interaction mean squares $MS(AB)$ is the appropriate denominator in the F ratio for testing the primary null hypothesis $H_0 : \alpha_i = 0$, whereas with fixed effects it was a substitute for the unavailable $MS(E)$. With fixed effects the use of $MS(AB)$ in the denominator was questionable, but for mixed effects it is the denominator we want.

A single missing value in an otherwise balanced design with $n = 1$ could still be estimated by (4.18).

For maximum likelihood estimation in the mixed model see Szatrowski and Miller (1980) and the references contained therein.

Multiple comparisons among the α_i can be handled as well under the mixed effects model. The only difference from fixed effects is that $MS(AB)$ is used as the estimate of σ^2. In particular, the *Tukey intervals* for a balanced design are

$$\alpha_i - \alpha_{i'} \in \bar{y}_{i\cdot\cdot} - \bar{y}_{i'\cdot\cdot} \pm q^{\alpha}_{I,(I-1)(J-1)} \left(\frac{MS(AB)}{nJ} \right)^{1/2}, \qquad (4.38)$$

where $q^{\alpha}_{I,(I-1)(J-1)}$ is the upper 100α percentile of a studentized range distribution for I variables with $(I-1)(J-1)$ df. For a mildly unbalanced design the approximate $MS(AB)$ from Table 4.2 can be used in (4.38) with the harmonic mean n^* of (4.12) substituted for n and $\bar{y}^*_{i\cdot\cdot}$ for $\bar{y}_{i\cdot\cdot}$. The probability coverage $1 - \alpha$ for all the intervals (4.38) with $i \neq i'$ will deteriorate as the imbalance increases when the harmonic mean n^* is used (see Dunnett, 1980a).

For just a single confidence interval (4.38) can be calculated with $t_{(I-1)(J-1)}^{\alpha/2}\sqrt{2}$ in place of $q_{I,(I-1)(J-1)}^{\alpha}$. For a limited number K of comparisons the *Bonferroni intervals*, which utilize $t_{(I-1)(J-1)}^{\alpha/2K}\sqrt{2}$ instead of $q_{I,(I-1)(J-1)}^{\alpha}$, are available. For $K < \binom{I}{2}$ these intervals can be shorter than the Tukey intervals.

Simultaneous confidence intervals are also available for more general contrasts. The Scheffé or Tukey intervals [i.e., (3.14) or (3.15), respectively] can be computed with $MS(AB)$ for $\hat{\sigma}^2$, nJ for n, and $(I-1)(J-1)$ for the df.

For *monotone alternatives* the linear contrast methods of Sections 3.1.3 and 4.1.3 can be applied to the balanced, or approximately balanced, mixed model. The only difference from these earlier sections is that the variance estimate of $\sum_{i=1}^{I} c_i \bar{y}_{i\cdot\cdot}$ is $MS(AB)\sum_{i=1}^{I} c_i^2/nJ$ for balanced designs. For mildly unbalanced designs the variance of $\sum_{i=1}^{I} c_i \bar{y}_{i\cdot\cdot}^*$ is estimated by $MS(AB)\sum_{i=1}^{I} c_i^2/n^*J$ where $MS(AB)$ and n^* are given by Table 4.2 and (4.12), respectively.

Although attention usually centers on the fixed effects in a mixed model, estimating the *variance components* σ_b^2 and $\sigma_{\alpha b}^2$ may also be of interest in some circumstances. The method of moments estimators

$$\begin{aligned} \hat{\sigma}_b^2 &= \frac{MS(B) - MS(E)}{nI}, \\ \hat{\sigma}_{\alpha b}^2 &= \frac{MS(AB) - MS(E)}{n}, \end{aligned} \tag{4.39}$$

with n^* in place of n for nearly balanced designs, are reminiscent of Section 3.5.1 and are forerunners of Section 4.7. Normal theory methods, in particular the method of Satterthwaite (1946), could be applied to produce confidence intervals for σ_b^2 and $\sigma_{\alpha b}^2$, but these are not especially recommended because of their sensitivity to normality. The jackknife method (see Section 3.6.3) should provide more robust results. To apply the jackknife one would successively delete each of the Factor B levels (columns).

4.6. Departures from Assumptions.

Virtually nothing has been published on the effects of various departures from the underlying assumptions for the mixed effects model. What we are to believe must be inferred from the known results on fixed effects and random effects models.

For *nonnormal* e_{ijk}, αb_{ij}, and b_j the effects on tests about the α_i should be minimal for balanced, or nearly balanced, designs. The block effects are eliminated, and the introduction of the random interactions does not change the character of the tests that much from the fixed effects case.

If appreciable nonnormality is present, there is not much that can be done about it. Perhaps a transformation will improve the analysis. The common rank tests require that no interactions be present. If this is the case, then the nonparametric tests described for fixed effects can be applied. For details on these corrective procedures the reader should consult Section 4.2.3.

The situation is different for the effects of nonnormality on tests and confidence intervals for σ_e^2, $\sigma_{\alpha b}^2$, and σ_b^2. Here the analysis can be led into catastrophic errors. Escape is through more robust procedures such as the jackknife. For details see Sections 3.6, 4.8, and Chapter 7.

What about *unequal variances*? This could occur either in the e_{ij}, αb_{ij}, or b_j variables. For tests on α_i the effect of different σ_e^2 and $\sigma_{\alpha b}^2$ on the analysis should be very similar to the fixed effects case with a single observation per cell studied by Box (1954b) since the interaction sum of squares is used in the denominator of the F statistic. The effects of different σ_e^2 on testing $H_0 : \sigma_{\alpha b}^2 = 0$ and $H_0 : \sigma_b^2 = 0$ should be similar to the one-way classification (see Sections 3.3 and 3.7) since $MS(E)$ is used in the denominators of the associated F statistics.

For substantially unequal variances about the only hope of correction is to find a transformation that stabilizes the variances and does not destroy the model (see Section 3.3.3).

No particularly insightful comments can be made on *dependence* within the e_{ij}, ab_{ij}, and b_j in the mixed effects model. The reader may wish to read, or reread, the discussion in Section 4.4.

RANDOM EFFECTS

4.7. Normal Theory.

The random effects model is not fraught with questions about assumptions as is the mixed effects model. Very simply, it is

$$y_{ijk} = \mu + a_i + b_j + ab_{ij} + e_{ijk}, \tag{4.40}$$

for $i = 1, \cdots, I$, $j = 1, \cdots, J$, $k = 1, \cdots, n_{ij}$, where the random components are distributed as

$$\begin{aligned} a_i &\quad \text{independent} \quad N(0, \sigma_a^2), \\ b_j &\quad \text{independent} \quad N(0, \sigma_b^2), \\ ab_{ij} &\quad \text{independent} \quad N(0, \sigma_{ab}^2), \\ e_{ijk} &\quad \text{independent} \quad N(0, \sigma_e^2), \end{aligned} \tag{4.41}$$

with independence between the different lettered variables.

Concerns have been expressed over the reasonableness of assuming that the interaction term ab_{ij} is tossed into the model independently of a_i and b_j. However, uncorrelatedness, which with normality becomes independence, does seem to emerge from finite sampling models that define the interaction to be a function of the main A and B effects. For details the reader is referred to Scheffé (1959, Section 7.4) and Cornfield and Tukey (1956).

The problem usually of interest is to estimate the *components of variance* σ_a^2, σ_b^2, σ_{ab}^2, and σ_e^2. However, on some rare occasions estimates of the individual components a_i, b_j, and ab_{ij} may be desired. These two problems are treated in the order cited.

The model (4.40) is referred to as a *cross-classification model.* A slightly different and equally important model is the *nested model.* For this latter model see (4.44) and the related discussion.

4.7.1. Estimation of Variance Components

The standard *method of moments* estimators for a balanced design (i.e., $n_{ij} \equiv n$) are based on the expected mean squares for the sums of square appering in Table 4.1. These expectations are

$$\begin{aligned} E[MS(A)] &= \sigma_e^2 + n\sigma_{ab}^2 + nJ\sigma_a^2, \\ E[MS(B)] &= \sigma_e^2 + n\sigma_{ab}^2 + nI\sigma_b^2, \\ E[MS(AB)] &= \sigma_e^2 + n\sigma_{ab}^2, \\ E[MS(E)] &= \sigma_e^2, \end{aligned} \tag{4.42}$$

so the associated estimators are

$$\begin{aligned} \hat{\sigma}_a^2 &= \frac{MS(A) - MS(AB)}{nJ}, \\ \hat{\sigma}_b^2 &= \frac{MS(B) - MS(AB)}{nI}, \\ \hat{\sigma}_{ab}^2 &= \frac{MS(AB) - MS(E)}{n}, \\ \hat{\sigma}_e^2 &= MS(E). \end{aligned} \tag{4.43}$$

The credentials of the estimators (4.43) are that they are uniform minimum variance unbiased estimators (UMVUE) under normal theory, and uniform minimum variance quadratic unbiased estimators (UMVQUE) in general; see Searle (1971, pp. 405–406) for additional discussion and references. They do, however, suffer the embarrassment of sometimes being negative, except for $\hat{\sigma}_e$ which is always positive. The actual maximum likelihood estimators would occur on a boundary rather than being negative (see Herbach, 1959). Personally, I would always adjust an estimate to zero rather than report a negative value.

It should certainly be possible to construct improved estimators along the lines of the Klotz-Milton-Zacks [see (3.55)] estimators used in the one-way classification. However, the details on these estimators have not been worked out by anyone for the two-way classification.

Similarly, it should be possible to construct formal Bayes estimators, but the details have not been worked out for the two-way classification. For discussion and references on Bayes estimators in the one-way classification see Klotz, Milton, and Zacks (1969), Portnoy (1971), and Searle (1971, p. 408).

See C. R. Rao (1970, 1971, 1972) for MINQUE estimation.

LaMotte (1973), Pukelsheim (1981), and others have investigated nonnegative unbiased variance component estimators.

An approximately balanced design can be handled by the preceding approach with the Table 4.2 approximate ANOVA replacing Table 4.1. On the other hand, extremely unbalanced designs are a horror story. A number of different methods have been proposed for handling them, but all involve extensive algebraic manipulations. The technical detail required to carry out these analyses exceeds the limitations set for this book so the reader is referred to the best exposition of this area, namely, Searle (1971, Chapters 10 and 11). I have not had any experience with the different methods discussed by Searle so I cannot recommend any one over another.

On occasion Factors A and B are such that it makes no sense to postulate the existence of interactions so the terms ab_{ij} should be dropped from (4.40). In this case σ_{ab}^2 disappears from (4.42), and the estimators for σ_a^2 and σ_b^2 in (4.43) can use $MS(E)$ in place of $MS(AB)$. Alternatively, one can add $SS(AB)$ and $SS(E)$ and divide by $IJ(n-1)+(I-1)(J-1)$ to form a combined estimate of σ_e^2. This combined estimate can then be substituted for $MS(AB)$ in the expressions for $\hat{\sigma}_a^2$ and $\hat{\sigma}_b^2$ in (4.43).

Another variation on the model (4.40) gives rise to the *nested model*. In my experience the nested model for components of variance problems occurs more frequently in practice than does the cross-classification model. In the nested model the main effects for one factor, say, B, are missing in (4.40). The reason is that the entities creating the different levels of Factor B are not the same for different levels of Factor A. For example, the levels (subscript i) of Factor A might represent different litters, and the levels (subscript j) of Factor B might be different animals, which are a different set for each litter. The additional subscript k might denote repeated measurements on each animal.

To be specific, the formal model for the nested design is

$$y_{ijk} = \mu + a_i + b_{ij} + e_{ijk}, \tag{4.44}$$

with

$$\begin{aligned} a_i &\quad \text{independent} \quad N(0, \sigma_a^2), \\ b_{ij} &\quad \text{independent} \quad N(0, \sigma_b^2), \\ e_{ijk} &\quad \text{independent} \quad N(0, \sigma_e^2), \end{aligned} \tag{4.45}$$

and independence between the different lettered variables. It is customary with this model to use the symbol b rather than ab because the interpretation for this term has changed from synergism or interaction to one of a main effect nested inside another main effect.

For a balanced design the *method of moments* estimators are based on the sums of squares

$$
\begin{aligned}
SS(A) &= nJ\sum_{i=1}^{I}(\bar{y}_{i\cdot\cdot} - \bar{y}_{\cdot\cdot\cdot})^2, \\
SS(B) &= n\sum_{i=1}^{I}\sum_{j=1}^{J}(\bar{y}_{ij\cdot} - \bar{y}_{i\cdot\cdot})^2, \\
SS(E) &= \sum_{i=1}^{I}\sum_{j=1}^{J}\sum_{k=1}^{K}(y_{ijk} - \bar{y}_{ij\cdot})^2,
\end{aligned}
\tag{4.46}
$$

which have degrees of freedom $I-1$, $I(J-1)$, and $IJ(n-1)$, respectively. The mean squares corresponding to (4.46) have the expectations

$$
\begin{aligned}
E(MS(A)) &= \sigma_e^2 + n\sigma_b^2 + nJ\sigma_a^2, \\
E(MS(B)) &= \sigma_e^2 + n\sigma_b^2, \\
E(MS(E)) &= \sigma_e^2,
\end{aligned}
\tag{4.47}
$$

so the estimators are

$$
\begin{aligned}
\hat{\sigma}_a^2 &= \frac{1}{nJ}(MS(A) - MS(B)), \\
\hat{\sigma}_b^2 &= \frac{1}{n}(MS(B) - MS(E)), \\
\hat{\sigma}_e^2 &= MS(E).
\end{aligned}
\tag{4.48}
$$

The increasing tier phenomenon exhibited in (4.47) holds for nested designs with more than two effects. The only complication arises when one or more of the estimates are negative. This is an indication that the corresponding variance components are zero or negligible. One might want to reset any negative estimates to zero, combine the adjacent sums of squares, and subtract the combined mean squares from the mean squares higher in the tier.

Extension of these ideas to the unbalanced design does not represent as formidable a task for the nested design as it does for the crossed design. The details for the case of two factors are given explicitly in Graybill (1961, pp. 354–359) and Searle (1971, pp. 475–476). The sums of squares (4.46), appropriately modified for unbalanced designs, form the basis for the analysis. It is even possible to allow for varying numbers J_i of levels of Factor B for different levels of Factor A.

4.7.2. Tests for Variance Components

Under normal theory the distributions of the sums of squares appearing in Table 4.1 are rather easy to derive and describe:

$$\begin{aligned} SS(M) &\sim (\sigma_e^2 + n\sigma_{ab}^2 + nI\sigma_b^2 + nJ\sigma_a^2)\cdot \\ &\qquad \chi_1^2\left(\frac{nIJ\mu^2}{\sigma_e^2 + n\sigma_{ab}^2 + nI\sigma_b^2 + nJ\sigma_a^2}\right), \\ SS(A) &\sim (\sigma_e^2 + n\sigma_{ab}^2 + nJ\sigma_a^2)\chi_{I-1}^2, \\ SS(B) &\sim (\sigma_e^2 + n\sigma_{ab}^2 + nI\sigma_b^2)\chi_{J-1}^2, \\ SS(AB) &\sim (\sigma_e^2 + n\sigma_{ab}^2)\chi_{(I-1)(J-1)}^2, \\ SS(E) &\sim \sigma_e^2\chi_{IJ(n-1)}^2, \end{aligned} \tag{4.49}$$

and all five sums of squares are independent.

To test the hypothesis $H_0 : \sigma_{ab}^2 = 0$, one uses the F ratio $MS(AB)/MS(E)$. To test $H_0 : \sigma_a^2 = 0$, one usually uses the F ratio $MS(A)/MS(AB)$, unless a decision has been made to combine $SS(AB)$ and $SS(E)$ in the denominator because σ_{ab}^2 is believed to be zero. An analogous F statistic provides a test for $H_0 : \sigma_b^2 = 0$. Under the alternative nonnull hypotheses, these ratios are distributed as the appropriate ratios of multiplicative constants from (4.49) times central F random variables. (For details see Secton 3.5.2.) Thus power calculations are made from central F tables in contrast to noncentral F tables for fixed effects models.

The F tests of $H_0 : \sigma_{ab}^2 = 0$ and $H_0 : \sigma_a^2 = 0$ mentioned in the preceding paragraph are uniformly most powerful similar tests.

However, they are not likelihood ratio tests, which are more complicated because of boundaries to the parameter space. For details and proofs of these assertions the reader is referred to Herbach (1959) and Gautschi (1959).

Although their general use is not recommended because of their extreme sensitivity to nonnormality (see Section 4.8), confidence intervals can be constructed based on the distribution theory (4.49). For σ_e^2 a confidence interval can be derived from $SS(E)/\sigma_e^2 \sim X_{IJ(n-1)}^2$. Similarly, confidence intervals on ratios of particular combinations of variance components can be obtained by taking the appropriate ratios of mean sums of squares from (4.49) as, for example,

$$\frac{MS(A)}{MS(AB)} \times \frac{\sigma_e^2 + n\sigma_{ab}^2}{\sigma_e^2 + n\sigma_{ab}^2 + nJ\sigma_a^2} \sim F_{I-1,(I-1)(J-1)}. \tag{4.50}$$

However, the problem of calculating confidence intervals for σ_{ab}^2, σ_b^2, and σ_a^2 separately is far more difficult. The complicated method of Bulmer (1957), which is described in Scheffé (1959, pp. 231–235), is available. However, the approximate method of Satterthwaite (1946) may produce just as good results. The idea behind this method was described in Sections 2.3.3, "Other Tests," and 3.5.2, and it easily extends to the two-way classification.

The nearly balanced design can be handled by the usual dodge of inserting n^* for n (see Table 4.2), but tests and confidence intervals for poorly balanced designs constitute a wasteland.

The distribution theory for the sums of squares (4.46) used in conjunction with nested designs is straightforward and simple:

$$\begin{aligned} SS(A) &\sim (\sigma_e^2 + n\sigma_b^2 + nJ\sigma_a^2)\chi_{I-1}^2, \\ SS(B) &\sim (\sigma_e^2 + n\sigma_b^2)\chi_{I(J-1)}^2, \\ SS(E) &\sim \sigma_e^2\chi_{IJ(n-1)}^2, \end{aligned} \tag{4.51}$$

and all three sums of squares are independent.

To test the hypothesis $H_0 : \sigma_b^2 = 0$ one uses the F ratio $MS(B)/MS(E)$, and to test $H_0 : \sigma_a^2 = 0$ the appropriate ratio is $MS(A)/MS(B)$. In all nested designs the higher line in the tier is always tested against the next lower line. If a conclusion is reached that $\sigma_b^2 = 0$, then the test of $H_0 : \sigma_a^2 = 0$ could be improved by combining $SS(B)$ and $SS(E)$ to form a denominator sum of squares with $I(J-1) + IJ(n-1)$ degrees of freedom. Under alternative hypotheses these F ratios are distributed as central F ratios multiplied by the appropriate ratio of variances. This can be exploited to produce confidence intervals on some variance ratios. However, one still needs to rely on the approximate Satterthwaite (1946) approach for constructing intervals on individual components (see Sections 2.3.3, "Other Tests," and 3.5.2).

4.7.3. Estimation of Individual Effects and Overall Mean

For the two-way crossed classification with random effects interest almost always is focused on estimating and testing σ_e^2, σ_{ab}^2, σ_b^2, and σ_a^2. However, it is not inconceivable that in some cases there might be interest as well, or instead, in estimating the cell means $\mu_{ij} = \mu + a_i + b_j + ab_{ij}$.

The classical approach would be to use the estimates $\hat{\mu}_{ij} = \bar{y}_{ij\cdot}$. However, viewed as a collection of estimates, one could do better (in terms of mean squared error) through the James-Stein (1961) and Lindley (1962) approach. The idea would be to shrink the individual estimates toward the common mean as in

$$\hat{\mu}_{ij} = \bar{y}_{\cdots} + (1 - S)(\bar{y}_{ij\cdot} - \bar{y}_{\cdots}), \tag{4.52}$$

where the shrinking factor S depends on the sums of squares $SS(E)$, $SS(AB)$, $SS(B)$, and $SS(A)$. Unfortunately, the specific details on the construction of an appropriate S have not been worked out for the two-way classification as they have been for the one-way classification (see Section 3.5.3).

Alternatively, attention might center on estimating $a_1, \cdots, a_I$, or, equivalently, on the levels of Factor B. Again, specific estimators have not been proposed to date for handling this situation.

In the nested design one sometimes wants an estimate and confidence interval for μ. One typically uses $\hat{\mu} = \bar{y}_{\cdots}$. In the balanced case this estimator has variance

$$\frac{\sigma_e^2}{IJn} + \frac{\sigma_b^2}{IJ} + \frac{\sigma_a^2}{I}. \tag{4.53}$$

This can be estimated by $MS(A)/IJn$. In the unbalanced case an estimate for the variability of $\bar{y}_{\cdots}$ can be obtained by substituting estimates $\hat{\sigma}_e^2$, $\hat{\sigma}_b^2$, and $\hat{\sigma}_a^2$ into the expression for the variance of $\bar{y}_{\cdots}$. Alternative estimators using different weights may be worth considering in the unbalanced case. For a pertinent discussion see Section 3.5.4.

4.8. Departures from Assumptions.

The effects of *nonnormality* in any of the sets of underlying random variables $\{e_{ijk}\}$, $\{ab_{ij}\}$, $\{b_j\}$, $\{a_i\}$ can be devastating to the distribution theory for the sums of squares involving them. The kurtoses of these variables have a substantial impact on the variances of the sums of squares. Confidence intervals, even those based on Satterthwaite's approximation, are not to be trusted. Tests on σ_e^2 are also very sensitive to nonnormality. The exceptions to this general nonrobustness are the variance ratio tests for the presence of a variance component, such as

$$\frac{MS(A)}{MS(AB)} \lessgtr F^{\alpha}_{I-1,(I-1)(J-1)} \tag{4.54}$$

for $H_0 : \sigma_a^2 = 0$. The denominator is converging by the law of large numbers to the correct normalizing constant under the null hypothesis. Also, under the null hypothesis, the variables corresponding to the component being tested are not present and the averaging over the other variables in the various row, column, and cell means dampens the effects of the kurtoses in the numerator. The larger the design, the better off one is in this regard.

For a fuller discussion of the effects of nonnormality on the distribution of variance estimates see Section 7.2.

Detection of nonnormality in anything but the e_{ijk} is usually hopeless. The reason is that, unless I and/or J are awfully large, there are just too few $\hat{a}_i = \bar{y}_{i\cdot\cdot} - \bar{y}_{\cdot\cdot\cdot}$, $\hat{b}_j = \bar{y}_{\cdot j\cdot} - \bar{y}_{\cdot\cdot\cdot}$, or $\widehat{ab}_{ij} = \bar{y}_{ij\cdot} - \bar{y}_{i\cdot\cdot} - \bar{y}_{\cdot j\cdot} + \bar{y}_{\cdot\cdot\cdot}$ to infer anything. In addition, $\hat{a}_i$ and $\hat{b}_j$ contain mixtures of the interactions ab_{ij} so that an uncontaminated view of a_i and b_j is impossible. There may be enough residuals $r_{ijk} = y_{ijk} - \bar{y}_{ij\cdot}$ to spot nonnormality in e_{ijk} through a combined probit plot (see Section 4.2.2).

There are no outstanding suggestions for how to cope with nonnormality in the random effects two-way classification. Possibly, a fortuitious transformation could be uncovered. For balanced designs application of the jackknife may be feasible (see Section 3.6.3).

So little is known about the effects of *unequal variances* and *dependence* on the random effects analysis in the two-way classification that no discussion is possible. Techniques for detection and correction of these assumption failures are nonexistent, except for what can be carried over from simpler designs.

Exercises.

1. Verify the expectations in (4.9) for the two-way crossed fixed effects model.

2. Verify the distribution theory for $SS(A)$ and $SS(E)$ stated in (4.8) for the two-way crossed fixed effects model.

3. For a balanced two-way classification with $n = 1$, prove that Tukey's SS for nonadditivitity

$$SS_1 = \frac{\left[\sum_{i=1}^{I} \sum_{j=1}^{J} \hat{\alpha}_i \hat{\beta}_j \widehat{\alpha\beta}_{ij}\right]^2}{\sum_{i=1}^{I} \hat{\alpha}_i^2 \sum_{j=1}^{J} \hat{\beta}_j^2}$$

is distributed as $\sigma^2 \chi_1^2$ under $H_0 : \alpha\beta_{ij} \equiv 0$.

Hint: Condition on $\{\hat{\alpha}_i\}$ and $\{\hat{\beta}_j\}$. Use the independence of $\{\widehat{\alpha\beta}_{ij}\}$ from $\{\hat{\alpha}_i\}$ and $\{\hat{\beta}_j\}$.

4. For a two-way crossed mixed model with I (fixed) rows, J (random) columns, and n replictions per cell, show that

$$E(MS(A)) = \sigma_e^2 + n\sigma_{\alpha b}^2 + \frac{nJ\sum_{i=1}^{I} \alpha_i^2}{I-1}.$$

5. For a two-way crossed mixed model with I (fixed) rows, J (random) columns, and n replications per cell, show that under the Graybill model (4.33)

$$SS(A) \sim (\sigma_e^2 + n\sigma_{\alpha b}^2)\chi_{I-1}^2 \left(\frac{nJ\sum_{i=1}^{I} \alpha_i^2}{\sigma_e^2 + n\sigma_{\alpha b}^2}\right).$$

Hint: Show that $\{\bar{y}_{i\cdot\cdot} - \bar{y}_{\cdot\cdot\cdot}\}$ has the same covariance structure as $\{z_i - \bar{z}\}$, where the z_i, $i = 1, \cdots I$, are independently, normally distributed with equal variances.

6. For a two-way nested mixed model, i.e.,

$$y_{ijk} = \mu + \alpha_i + b_{ij} + e_{ijk},$$

with $i = 1, \cdots, I$, $j = 1, \cdots, J$, $k = 1, \cdots, n > 1$, $\sum_{i=1}^{I} \alpha_i = 0$, and

$$\begin{array}{lll} b_{ij} & \text{independent} & N(0, \sigma_b^2), \\ e_{ijk} & \text{independent} & N(0, \sigma_e^2), \\ \{b_{ij}\} & \text{and} \quad \{e_{ijk}\} & \text{independent}, \end{array}$$

construct a test of $H_0 : \alpha_i \equiv 0, i = 1, \cdots, I$, vs. $H_1 : \alpha_i \not\equiv 0$.

Note: $\{\alpha_i\}$ might be different treatment effects, $\{b_{ij}\}$ might be subject effects with different subjects for each treatment, and $\{e_{ijk}\}$ might be repeated measurements on each subject.

7. Verify the expectations in (4.47) for the two-way nested random effects model.

8. Verify the distribution theory stated in (4.51) for the two-way nested random effects model.

9. In a study of platelet production, 40 rats were equally separated into altitude chambers, the experimental group at 15,000 ft. and the control group at sea level. Half of the rats were splenectomized (i.e., spleen removed), and the other half were nonsplenectomized. Various blood parameters were measured over a succession of days.* The fibrinogen levels (in mg%) on day 21 are reported in the table. Some data are missing.

   Determine if there are significant effects due to altitude and splenectomy.

---

* Rand, K. H., Anderson, T., Lukis, G. A., and Creger, W. P. (1970). Effect of hypoxia on platelet level in the rat. *Clinical Research*, **18**, 178 (abstract).

	Splenectomy	
	Yes	No
Altitude	528	434
	444	331
	338	312
	342	575
	338	472
	331	444
	288	575
	319	384
Control	294	272
	254	275
	352	350
	241	350
	291	466
	175	388
	241	425
	238	344
	269	425

10. The ability of radiologists to visualize vascular structures has progressed through the development of contrast agents and radiographic imaging technology. Using digital subtraction angiography with measurements from a modified CT scanner, a Stanford study compared 6 contrast agents injected sequentially into the arteries of dogs at 10 minute intervals. Although this time interval was considered sufficient to eliminate any residual effect from a previous injection, an extra period Latin square design with 6 dogs was used to permit statistical testing for residual effects as well as main effects. The design and the values for the opacification index computed from photon absorption are

displayed in the table.

A full analysis indicated no period and residual effects.* In your analysis discard the data from the extra period and assume no period effects. Test for differences in contrast agents (A, B, C, D, E, F).

	Dog					
Period	1	2	3	4	5	6
1	A 363	B 259	C 300	D 407	E 221	F 156
2	C 349	D 326	E 236	F 286	A 267	B 254
3	B 212	C 280	D 309	E 427	F 227	A 234
4	E 203	F 245	A 267	B 291	C 287	D 319
5	F 221	A 265	B 189	C 413	D 364	E 251
6	D 272	E 311	F 238	A 442	B 262	C 225
7	D 368	E 321	F 267	A 422	B 263	C 257

11. For the data in Exercise 11 of Chapter 3 allow for a nursing pair effect, and estimate by the method of moments its variance component in addition to the components for experiments and error.

* Burbank, F. H., Brody, W. R., Hall, A., and Keyes, G. (1982). A quantitative in vivo comparison of six contrast agents by digital subtraction angiography. *Investigative Radiology*, **17**, 610–616. For technical details on the ANOVA for extra period Latin square designs see Lucas, H. L. (1957), Extra-period Latin-square change-over design, *Journal of Dairy Science*, **40**, 225–239, or Cochran, W. G. and Cox, G. M. (1957), *Experimental Designs*, Second Edition, Wiley, New York, Sec. 4.65a (pp. 139–141).

Chapter 5

REGRESSION

With each value of the variable y there may be associated the value of another variable x. Both variables may be of equal stature and interest, and the statistical problem is to investigate the relationship between them. In other instances, the variable x may be a baseline value against which the value of the primary variable y should be compared, or x may be an explanatory variable whose effect on the primary variable y should be adjusted for or standardized. The appropriate analyses for these situations are the topics of this chapter and the next.

A common statistical problem involves repeated measurements under different conditions. For example, x might be the pretreatment value for a patient and y the posttreatment value. Similar paired settings include studies of twins or measurements on the two arms (or legs) of each subject, where the two twins or extremities receive different treatments. In Chapter 1 it was suggested that the comparison between x and y be handled as a one sample problem by computing the difference $y - x$ for each pair. This is typically the appropriate approach, but it does assume that, except for random error, the x and y values lie on a line with slope 1 whose intercept is the average difference between the variables. This underlying model is displayed in Figure 5.1.

Another situation, somewhat less commonly encountered, is where the x and y values follow a ray emanating from the origin

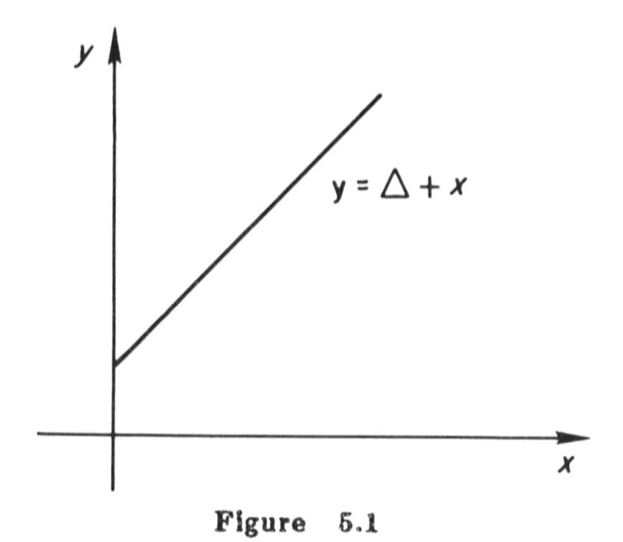

Figure 5.1

as in Figure 5.2. In the difference model of Figure 5.1, x and y are related by $y = \Delta + x$, where Δ is constant except for random variation, but in the model for Figure 5.2, $y = \rho x$, where ρ is constant except for randomness in the data.

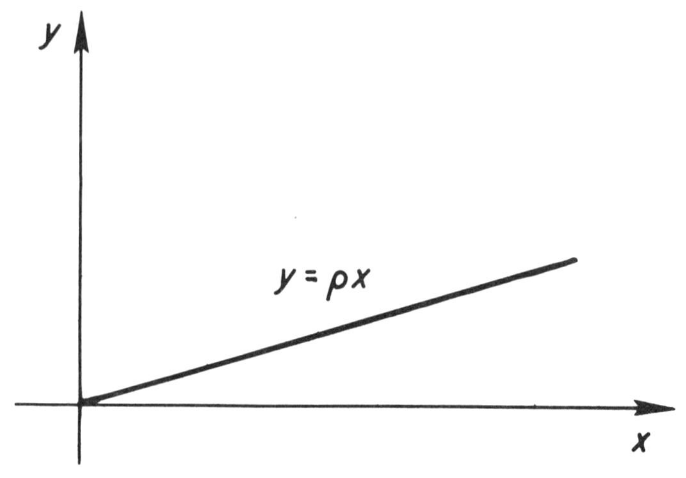

Figure 5.2

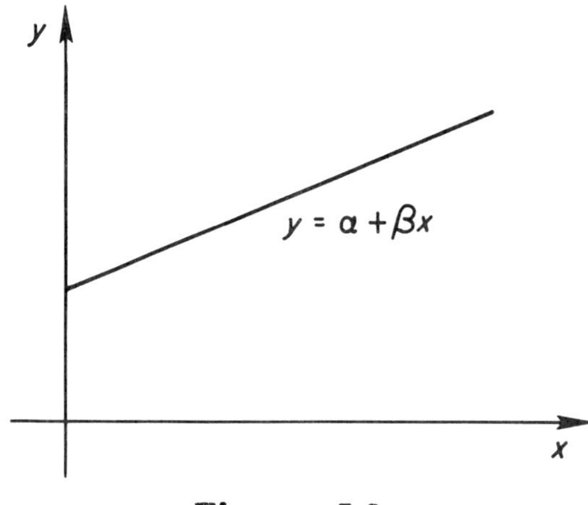

Figure 5.3

The ratio or multiplicative model $y = \rho x$ can be handled in one of the following ways. The first possibility is to take logarithms of x and y and use the methods of Chapter 1. This is appropriate if the errors look normal and homoscedastic after the log transformation. A second possibility is to use ratios, taken singly as y_i/x_i and then averaged or as the ratio of averages $\bar{y}/\bar{x}$. This is the subject of Chapter 6. The third possibility is to treat the analysis as a regression problem in which the intercept is known to be zero. This approach is considered in this chapter.

Figure 5.3 illustrates the most general linear relationship between x and y, namely $y = \alpha + \beta x$. Estimation and testing of the intercept α and the slope β lie in the domain of regression analysis, which is the topic of this chapter. Because of the increase in complexity of the analysis, it is the least preferable method for handling paired values, but at times it is unavoidable. Although the difference model $y = \Delta + x$ is often appropriate for such measurements as post vs. pre, and left vs. right, it is more typical for the full model $y = \alpha + \beta x$ to be required when x is an explanatory variable such as age or weight, and not the same measurement as y at a different time, site, etc.

In regression analysis, given the value of x, the variable y is assumed to fluctuate randomly about the central value $\alpha + \beta x$. The distribution theory of the estimates and tests takes the x values to be fixed. If, in fact, the x values are themselves random variables, the same analysis pertains, but it is now a conditonal analysis, conditioned on the observed values of the x variable.

If, in addition to inherent variability, the y variable is measured with inaccuracy, there is still no change in the analysis. The unexplained variability about the regression line simply has another component added to it. However if the x variable is measured with nontrivial error, the standard analysis should not be used because it leads to biased estimates. For this reason this chapter is divided into two parts. The first describes the standard *regression model* and analysis, and the second is devoted to the *errors-in-variables model*, which is a term often used for the situation where the x variable contains measurement error.

This chapter considers only the case of a single variable x linearly related to y. Polynomial regression and regression with more than one predictor variable are all topics in multiple regression, which is beyond the intended scope of this book. There are many excellent books on multiple regression, and I especially recommend Draper and Smith (1981).

REGRESSION MODEL

5.1. Normal Linear Model.

5.1.1. One Sample: General Intercept

The standard model assumes that the observations $y_1, \cdots y_n$ satisfy

$$y_i = \alpha + \beta x_i + e_i, \tag{5.1}$$

where the e_i are independently distributed as $N(0, \sigma^2)$. Although (5.1) is stated simply in terms of the observed x_i and y_i, the investigator usually has in mind that, given an x value, the variable y is normally distributed with

$$E(y \mid x) = \mu(x) = \alpha + \beta x \tag{5.2}$$

and

$$\text{Var}(y \mid x) = \sigma^2(x) \equiv \sigma^2. \tag{5.3}$$

The parameter α is the intercept on the y-axis when $x = 0$, and β is the slope of the regression line (5.2). The linear relationship (5.2) is assumed to hold over a range of x values, but this range may be limited. Assumption (5.3) requires the variance to be constant over this range.

No assumption has been stated about $x_1, \cdots, x_n$. They can be fixed values such as dosage levels in a bioassay or selected consecutive time points. At other times the x_i value may be whatever comes along with y_i. Examples of this are the age and weight of the subject or the ambient temperature at the time of measurement. In this latter context where the x_i may themselves be random, they are nonetheless thought of as being fixed in the analysis. The distribution theory is conditioned on the observed values of x; therefore, the resulting tests and confidence intervals are conditional ones.

In regression analysis with explanatory variables, the variable y is often referred to as the *dependent variable* and the variable x as the *independent variable*. This can be confusing to the novice because of the independence assumption on the y_i. Alternative terminology is to refer to y as the *response variable* and x as the *predictor variable*.

When x is a baseline or other explanatory variable, the usual statistical problem is to estimate α and β. Sometimes one also wants to test whether the intercept and slope equal certain preconceived values, such as 0 for α and 1 for β. Since the (conditional) mean of y depends on x, the problem on occasion is to estimate the mean of y at a standardized value x_0, i.e., $\mu(x_0) = \alpha + \beta x_0$. This is called the *prediction problem*, and the reverse of this is the *calibration problem*. In calibration the problem is to estimate the value x_0 for which $\mu(x_0)$ equals a specified value μ_0 (i.e., $x_0 = (\mu_0 - \alpha)/\beta$). This type of problem occurs frequently in bioassay.

When neither variable is subordinate, the problem is to investigate the relationship between x and y. If the investigator computes the correlation coefficient r between x and y, he or she is examining the extent of the linear relationship between x and y. In the case of bivariate normally distributed variables, there are two nonidentical linear regressions, namely,

$$E(y \mid x) = \alpha + \beta x, \tag{5.4}$$

and

$$E(x \mid y) = \alpha' + \beta' y. \tag{5.5}$$

These lines are distinctly different as indicated in Figure 5.4. However, testing whether the correlation is zero is equivalent to testing whether $\beta = 0$ and $\beta' = 0$.

The least squares and maximum likelihood estimates of α and

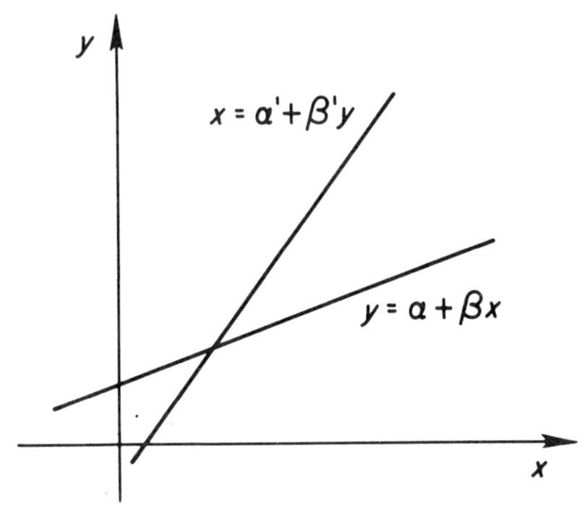

Figure 5.4

β are

$$\hat{\alpha} = \bar{y} - \hat{\beta}\bar{x},$$
$$\hat{\beta} = \frac{\sum_{i=1}^{n}(x_i - \bar{x})(y_i - \bar{y})}{\sum_{i=1}^{n}(x_i - \bar{x})^2}, \tag{5.6}$$

where the computational formulas

$$\sum_{i=1}^{n}(x_i - \bar{x})(y_i - \bar{y}) = \sum_{i=1}^{n} x_i y_i - n\bar{x}\bar{y},$$
$$\sum_{i=1}^{n}(x_i - \bar{x})^2 = \sum_{i=1}^{n} x_i^2 - n\bar{x}^2, \tag{5.7}$$

may be used. The bias-corrected maximum likelihood estimator of σ^2 is*

* The maximum likelihood estimate of σ^2 has the denominator n rather than $n-2$; the latter denominator makes the estimate unbiased.

$$\begin{aligned}
\hat{\sigma}^2 &= \frac{1}{n-2}\sum_{i=1}^{n}(y_i - \hat{\alpha} - \hat{\beta}x_i)^2, \\
&= \frac{1}{n-2}\left\{\sum_{i=1}^{n} y_i^2 - n\bar{y}^2 - \hat{\beta}^2 \sum_{i=1}^{n}(x_i - \bar{x})^2\right\}, \qquad (5.8)\\
&= \frac{1}{n-2}\left\{\sum_{i=1}^{n} y_i^2 - n\bar{y}^2 - \frac{\left[\sum_{i=1}^{n}(x_i - \bar{x})(y_i - \bar{y})\right]^2}{\sum_{i=1}^{n}(x_i - \bar{x})^2}\right\}.
\end{aligned}$$

For bivariate normally distributed variables (x, y), the maximum likelihood estimate of the correlation coefficient $\rho = \text{Cov}(x, y)$ $/SD(x)SD(y)$ is

$$r = \frac{\sum_{i=1}^{n}(x_i - \bar{x})(y_i - \bar{y})}{\left[\sum_{i=1}^{n}(x_i - \bar{x})^2 \sum_{i=1}^{n}(y_i - \bar{y})^2\right]^{1/2}}, \qquad (5.9)$$

where the formulas (5.7) may be used for computation. The estimator r is called the *product-moment correlation coefficient.*

The estimators $(\hat{\alpha}, \hat{\beta})$ have a bivariate normal distribution with means (α, β) and variances-covariance given by

$$\begin{aligned}
\text{Var}(\hat{\alpha}) &= \sigma^2\left(\frac{1}{n} + \frac{\bar{x}^2}{\sum_{i=1}^{n}(x_i - \bar{x})^2}\right), \\
\text{Cov}(\hat{\alpha}, \hat{\beta}) &= \sigma^2\left(\frac{-\bar{x}}{\sum_{i=1}^{n}(x_i - \bar{x})^2}\right), \qquad (5.10)\\
\text{Var}(\hat{\beta}) &= \sigma^2\left(\frac{1}{\sum_{i=1}^{n}(x_i - \bar{x})^2}\right).
\end{aligned}$$

The variance estimator $\hat{\sigma}^2$ is distributed independently of $(\hat{\alpha}, \hat{\beta})$ and has a scaled χ^2 distribution with $n - 2$ df, i.e.,

$$\frac{(n-2)\hat{\sigma}^2}{\sigma^2} \sim \chi^2_{n-2}. \qquad (5.11)$$

If the model were recast as

$$y_i = \alpha^* + \beta^*(x_i - \bar{x}) + e_i, \tag{5.12}$$

where $\beta^* = \beta$ and $\alpha^* = \alpha + \beta\bar{x}$, then the least squares and maximum likelihood estimates

$$\hat{\alpha}^* = \bar{y}, \quad \hat{\beta}^* = \hat{\beta} \tag{5.13}$$

would be independently and normally distributed with means (α^*, β^*) and variances

$$\begin{aligned} \operatorname{Var}(\hat{\alpha}^*) &= \frac{\sigma^2}{n}, \\ \operatorname{Var}(\hat{\beta}^*) &= \frac{\sigma^2}{\sum_{i=1}^{n}(x_i - \bar{x})^2}. \end{aligned} \tag{5.14}$$

The models (5.1) and (5.12) are the same, but the independence of $\hat{\alpha}^*$ and $\hat{\beta}^*$ permits easier derivation of some expressions for tests and confidence intervals. The formula and the distribution theory for $\hat{\sigma}^2$ remain unchanged under this model formulation.

If one wants to test the null hypothesis that the intercept α has a specified α_0, as for example where α_0 is 0, the ratio

$$\frac{\hat{\alpha} - \alpha_0}{\hat{\sigma}\left(\frac{1}{n} + \frac{\bar{x}^2}{\sum_{i=1}^{n}(x_i - \bar{x})^2}\right)^{1/2}} \tag{5.15}$$

has a t distribution with $n - 2$ df under $H_0 : \alpha = \alpha_0$. One-sided or two-sided P values for the observed value of the ratio can be calculated from t tables or computer routines for the t distribution. The corresponding $100(1 - \alpha)\%$ confidence interval for α is

$$\alpha \in \hat{\alpha} \pm t_{n-2}^{\alpha/2}\hat{\sigma}\left(\frac{1}{n} + \frac{\bar{x}^2}{\sum_{i=1}^{n}(x_i - \bar{x})^2}\right)^{1/2}, \tag{5.16}$$

where $t_{n-2}^{\alpha/2}$ is the upper $100(\alpha/2)$ percentile of the t distribution with $n - 2$ df.

The ratio

$$\frac{\hat{\beta} - \beta_0}{\hat{\sigma}\sqrt{1/\sum_{i=1}^{n}(x_i - \bar{x})^2}} \tag{5.17}$$

is used to test that the slope β has a specified value β_0, as for example where β_0 is 1. Under $H_0 : \beta = \beta_0$, the ratio (5.17) has a t distribution with $n - 2$ df. The corresponding confidence interval is

$$\beta \in \hat{\beta} \pm t_{n-2}^{\alpha/2}\hat{\sigma}\left(\frac{1}{\sum_{i=1}^{n}(x_i - \bar{x})^2}\right)^{1/2}. \tag{5.18}$$

Sometimes one wants to test a joint null hypothesis $H_0 : \alpha = \alpha_0$, $\beta = \beta_0$, which amounts to specifying the line $\alpha_0 + \beta_0 x$. An example might be the 45° line with $\alpha_0 = 0$ and $\beta_0 = 1$. The classical test is to compare the value of $(\hat{\boldsymbol{\beta}} - \boldsymbol{\beta}_0)^T \boldsymbol{\Sigma}^{-1}(\hat{\boldsymbol{\beta}} - \boldsymbol{\beta}_0)/2\hat{\sigma}^2$ with the critical points of an F distribution on 2 and $n - 2$ df, where $\hat{\boldsymbol{\beta}} = (\hat{\alpha},\ \hat{\beta})^T$, $\boldsymbol{\beta}_0 = (\alpha_0, \beta_0)^T$, and $\boldsymbol{\Sigma}$ is the covariance matrix for $\hat{\boldsymbol{\beta}}$ given by (5.10) without the scalar multiple σ^2. Joint confidence intervals for α and β can be obtained by projecting the confidence ellipsoid for $\boldsymbol{\beta} = (\alpha, \beta)^T$ generated by the equation $(\hat{\boldsymbol{\beta}} - \boldsymbol{\beta})^T \boldsymbol{\Sigma}^{-1}(\hat{\boldsymbol{\beta}} - \boldsymbol{\beta})/2\hat{\sigma}^2 = F_{2,n-2}^{\alpha}$ onto the coordinate axes (see Scheffé intervals in Section 3.1.2 or Miller, 1981, pp. 58–60):

$$\begin{aligned} \alpha &\in \hat{\alpha} \pm (2F_{2,n-2}^{\alpha})^{1/2}\hat{\sigma}\left(\frac{1}{n} + \frac{\bar{x}^2}{\sum_{i=1}^{n}(x_i - \bar{x})^2}\right)^{1/2}, \\ \beta &\in \hat{\beta} \pm (2F_{2,n-2}^{\alpha})^{1/2}\hat{\sigma}\left(\frac{1}{\sum_{i=1}^{n}(x_i - \bar{x})^2}\right)^{1/2}. \end{aligned} \tag{5.19}$$

Shorter intervals are obtained by substituting the Bonferroni critical constant $t_{n-2}^{\alpha/4}$ for $(2F_{2,n-2}^{\alpha})^{1/2}$ in (5.19); see Section 3.1.2. With either critical constant, the intervals (5.19) have a probability exceeding $1 - \alpha$ of jointly containing the true paramter values α and β.

If one is willing to have the test statistic and confidence intervals expressed in terms of the reformulated, but equivalent, null hypothesis $H_0 : \alpha^* = \alpha_0^*$, $\beta^* = \beta_0^*$, where $\alpha_0^* = \alpha_0 + \beta_0 \bar{x}$, the computations simplify and the confidence intervals shorten. The classical test statistic is now

$$\left[n(\hat{\alpha}^* - \alpha_0^*)^2 + \left(\sum_{i=1}^{n} (x_i - \bar{x})^2 \right) (\hat{\beta}^* - \beta^*)^2 \right] \Big/ 2\hat{\sigma}^2, \tag{5.20}$$

which has an F distribution with 2 df for the numerator and $n - 2$ df for the denominator under H_0. However, the shortest confidence intervals, which have probability exactly equal to $1 - \alpha$ of containing α^* and β^*, are

$$\begin{aligned} \alpha^* &\in \hat{\alpha}^* \pm |m|_{2,n-2}^{\alpha} \hat{\sigma} \left(\frac{1}{n} \right)^{1/2}, \\ \beta^* &\in \hat{\beta}^* \pm |m|_{2,n-2}^{\alpha} \hat{\sigma} \left(\frac{1}{\sum_{i=1}^{n} (x_i - \bar{x})^2} \right)^{1/2}, \end{aligned} \tag{5.21}$$

where $|m|_{2,n-2}^{\alpha}$ is the upper 100α percentile of the studentized maximum modulus distribution with two independent variables in the numerator and $n - 2$ df for the denominator.* Good tables of $|m|_{2,n-2}^{\alpha}$ are available in Hahn and Hendrickson (1971), and these are reproduced in Miller (1981). The interval for $\beta = \beta^*$ in (5.21) is the same as in (5.19) except that $|m|_{2,n-2}^{\alpha}$ is smaller than $(2F_{2,n-2}^{\alpha})^{1/2}$ and $t_{n-2}^{\alpha/4}$.** The interval for α^* in (5.21) amounts to a confidence interval for the value of the regression line $\alpha + \beta x$ at $x = \bar{x}$, whereas the interval in (5.19) is a confidence interval for the regression line

* A studentized maximum modulus variable $|m|_{k,\nu}$ is distributed as $\max\{|y_1|, \cdots, |y_k|\}/(\chi_\nu^2/\nu)^{1/2}$, where $y_1, \cdots, y_k$ are independent $N(0,1)$, χ_ν^2 has a χ^2 distribution with ν df, and χ_ν^2 and $y_1, \cdots, y_k$ are independent.

** The critical constant $|m|_{2,n-2}^{\alpha}$ can be used in place of $(2F_{2,n-2}^{\alpha})^{1/2}$ in (5.19) as well. The probability of the intervals (5.19) with $|m|_{2,n-2}^{\alpha}$ jointly covering the true parameters is still greater than or equal to $1 - \alpha$. This follows from Šidák's inequality (see the first inequalty in Corollary 2 to Theorem 2 in Šidák, 1967). However, $|m|_{2,n-2}^{\alpha}$ is only slightly smaller than $t_{n-2}^{\alpha/4}$.

value at $x = 0$.

In general, the sample correlation coefficient (5.9) has a complicated distribution that depends on the parameter ρ (see T. W. Anderson, 1958, p. 69; C. R. Rao, 1973, p. 208, or other multivariate or general texts). However, one can verify algebraically that

$$\frac{r\sqrt{n-2}}{\sqrt{1-r^2}} = \frac{\hat{\beta}}{\hat{\sigma}\sqrt{1/\sum_{i=1}^{n}(x_i - \bar{x})^2}}. \tag{5.22}$$

The ratio (5.22) can be used to test the null hypothesis $H_0 : \rho = 0$ (and, equivalently, $H_0 : \beta = 0$ and $H_0 : \beta' = 0$) because under H_0 (5.22) has a t distribution with $n - 2$ df [see (5.17)]. For nonnull values of ρ the transformed correlation coefficient

$$\tanh^{-1} r = \frac{1}{2} \log\left(\frac{1+r}{1-r}\right) \tag{5.23}$$

has an asymptotic normal distribution with mean $\tanh^{-1}\rho + [\rho/2(n-1)]$ and variance $1/(n-3)$. Approximate tests and confidence intervals can be constructed with the aid of this transformation, which is due to R. A. Fisher (1921). (See Gayen, 1951, and Hotelling, 1953, for the correct moment expansions). Unfortunately, the asymptotic nonnull variance of (5.23) is sensitive to the assumption of normality. This makes confidence intervals based on this approach dangerous to use in practice indiscriminately (see Section 5.3).

On occasion the investigator may want to estimate the value of the regression function at a specified value x_0 of the independent variable and surround the estimate with a confidence interval. This is referred to as the *prediction problem*. The specified x_0 can be an interpolated value (i.e., within the range of $x_1, \cdots, x_n$) or an extrapolated value (i.e., outside the range of $x_1, \cdots, x_n$). In the case of extrapolation, the conformity of the regression function to a line over the extended range comes into question. The commonly used estimate for the regression value $\mu(x_0) = \alpha + \beta x_0$ is

$$\hat{\mu}(x_0) = \hat{\alpha} + \hat{\beta} x_0, \tag{5.24}$$

and the associated confidence interval is

$$\mu(x_0) \in \hat{\alpha} + \hat{\beta} x_0 \pm t_{n-2}^{\alpha/2} \hat{\sigma} \left(\frac{1}{n} + \frac{(x_0 - \bar{x})^2}{\sum_{i=1}^{n} (x_i - \bar{x})^2} \right)^{1/2}, \tag{5.25}$$

which is easily derived from (5.14).

On rare occasions there is interest in several different values for x_0 or possibly a continuous range of values. The estimates are the obvious ones obtained from $\hat{\mu}(x) = \hat{\alpha} + \hat{\beta} x$, but, with regard to confidence intervals, what is called for is a *confidence band* on the regression function. The first band to be proposed was the Working-Hotelling (1929) band:

$$\mu(x) \in \hat{\alpha} + \hat{\beta} x \pm (2F_{2,n-2}^{\alpha})^{1/2} \hat{\sigma} \left(\frac{1}{n} + \frac{(x - \bar{x})^2}{\sum_{i=1}^{n} (x_i - \bar{x})^2} \right)^{1/2}. \tag{5.26}$$

The probability that the intervals (5.26) are correct for all x between $-\infty$ and $+\infty$ is $1 - \alpha$ (see Miller, 1981, pp. 110–114). If the intervals for only a few x are used, then the probability exceeds $1 - \alpha$ somewhat.

Because the bands in (5.26) are hyperbolas (see Figure 5.5), they are time-consuming to calculate and draw by hand. For computer graphics this is not a problem. Easier bands to construct by hand are the straight-line bands (see Figure 5.6) of Graybill and Bowden (1967):

$$\mu(x) \in \hat{\alpha} + \hat{\beta} x \pm |m|_{2,n-2}^{\alpha} \hat{\sigma} \left(\frac{1}{\sqrt{n}} + \frac{|x - \bar{x}|}{\sqrt{\sum_{i=1}^{n} (x_i - \bar{x})^2}} \right), \tag{5.27}$$

where $|m|_{2,n-2}^{\alpha}$ is the upper 100α percentile of the studentized maximum modulus distribution with two independent variables in the

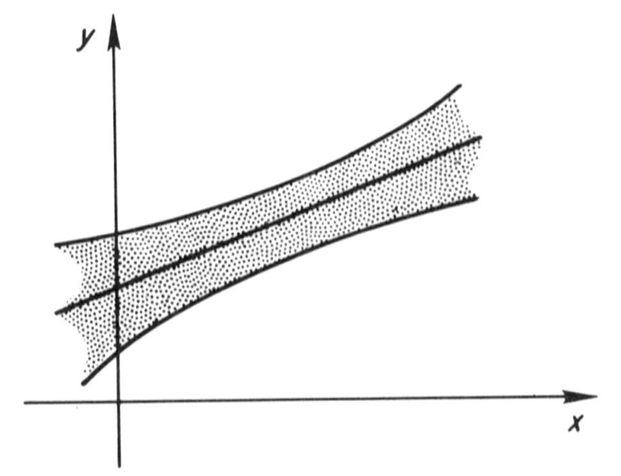

Figure 5.5

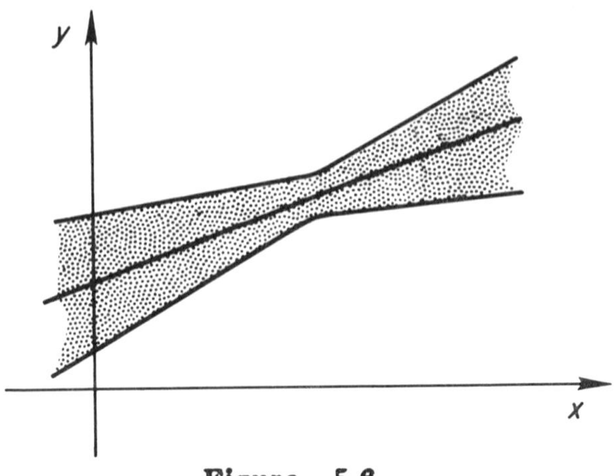

Figure 5.6

numerator and $n-2$ df in the denominator. See Hahn and Hendrickson (1971) for tables of $|m|^{\alpha}_{2,n-2}$. The intervals (5.27) are easily derived from (5.21). They are narrower than the intervals in (5.26) for x near $\bar{x}$ and $\pm\infty$, but they are somewhat wider for middling values of x between $\bar{x}$ and $\pm\infty$.

The reverse problem, which is called the *calibration problem*, is to decide which value of x leads to a specified value μ_0. From the regression line it follows that $x_0 = (\mu_0 - \alpha)/\beta$, so the standard estimate is

$$\hat{x}_0 = \frac{\mu_0 - \hat{\alpha}}{\hat{\beta}}. \tag{5.28}$$

This is a biased estimate due to $\hat{\beta}$ occurring in the denominator (see Chapter 6). Based on the second order term in the power series expansion

$$\frac{1}{\hat{\beta}} = \frac{1}{\beta} + (\hat{\beta} - \beta)\left(\frac{-1}{\beta^2}\right) + \frac{(\hat{\beta} - \beta)^2}{2}\left(\frac{2}{\beta^3}\right) + \cdots, \tag{5.29}$$

an adjusted estimate that reduces the order of the bias is

$$\hat{\hat{x}}_0 = \bar{x} + \left(\frac{\mu_0 - \bar{y}}{\hat{\beta}}\right)\left(1 - \frac{\hat{\sigma}^2}{\hat{\beta}^2 \sum_{i=1}^n (x_i - \bar{x})^2}\right). \tag{5.30}$$

A confidence interval for x_0 can be constructed by realizing that the ratio

$$\frac{(\hat{\alpha} + \hat{\beta}x_0 - \mu_0)^2}{\hat{\sigma}^2\left(\frac{1}{n} + \frac{(x_0 - \bar{x})^2}{\sum_{i=1}^n (x_i - \bar{x})^2}\right)} \tag{5.31}$$

has an F distribution with 1 and $n - 2$ df. Setting (5.31) equal to $F^{\alpha}_{1,n-2}$ and solving the resulting quadratic equation in x_0 for the two roots yields a confidence interval in most cases. The two roots are

$$\bar{x} + \frac{(\mu_0 - \bar{y}) \pm (F^{\alpha}_{1,n-2})^{1/2}\hat{\sigma}\left[\frac{1}{n}(1 - \epsilon) + \frac{(\mu_0 - \bar{y})^2}{\hat{\beta}^2 \sum_{i=1}^n (x_i - \bar{x})^2}\right]^{1/2}}{\hat{\beta}(1 - \epsilon)}, \tag{5.32}$$

where*

$$\epsilon = \frac{F^{\alpha}_{1,n-2}\hat{\sigma}^2}{\hat{\beta}^2 \sum_{i=1}^n (x_i - \bar{x})^2}. \tag{5.33}$$

---

* Note that $(F^{\alpha}_{1,n-2})^{1/2} = t^{\alpha/2}_{n-2}$.

However, the roots can be imaginary, in which case the confidence interval is the entire real line. Also, when $\hat{x}_0$ does not lie between the roots, the confidence region consists of the two infinite intervals above and below the two roots.

The aforementioned confidence interval procedure is credited to Fieller (1940, 1954). For greater detail the reader is referred to Chapter 6 or to Miller (1981, pp. 117–120) for discussion and figures on a closely related problem.

The ratio

$$\frac{\hat{\sigma}}{\hat{\beta}\,[\sum_{i=1}^{n}(x_i - \bar{x})^2]^{1/2}} \tag{5.34}$$

appearing in (5.30) and (5.33) is the estimated coefficient of variation of $\hat{\beta}$ [denoted by $\widehat{CV}(\hat{\beta})$]. When it is small (e.g., $< .10$), the regression slope $\hat{\beta}$ is accurately determined. In this case the bias correction in (5.30) is negligible (e.g., $< .02$) and can be ignored. Also, in this case ϵ in (5.33) is small (e.g., $< .05$), so the confidence interval

$$x_0 \in \hat{x}_0 \pm t_{n-2}^{\alpha/2}\frac{\hat{\sigma}}{|\hat{\beta}|}\left[\frac{1}{n} + \frac{(\mu_0 - \bar{y})^2}{\hat{\beta}^2\sum_{i=1}^{n}(x_i - \bar{x})^2}\right]^{1/2} \tag{5.35}$$

gives a good approximation to the fully exact interval (5.32). The factor multiplying $t_{n-2}^{\alpha/2}$ in (5.35) is the estimated standard deviation of $(\mu_0 - \bar{y})/\hat{\beta}$ obtained by the delta method (see Section 2.3.3 "Transformations"). For additional detail see Chapter 6 or Finney (1978, pp. 80–82).

It may be that several or many values of μ_0 (and corresponding x_0) are of interest, not just a single one. The confidence bands (5.26) or (5.27) can be used to construct confidence intervals for arbitrarily many values $x(\mu)$, that have probability at least $1 - \alpha$ of all being simultaneously correct. The procedure is to draw a horizontal line through the value μ on the vertical axis. The region of x values where the horizontal line is contained inside the band constitutes the confidence region for $x(\mu)$.

If the Working-Hotelling bands (5.26) are used, the confidence interval for $x(\mu)$ is given by (5.32) with $(2F^{\alpha}_{2,n-2})^{1/2}$ replacing $(F^{\alpha}_{1,n-2})^{1/2}$. Pathologies in the confidence region (i.e., the confidence interval is the entire real line or two infinite intervals) can occur just as for a single value x_0. However, for a small coefficient of variation $\widehat{CV}(\hat{\beta})$ [see (5.34)], this does not happen. For quite small $\widehat{CV}(\hat{\beta})$ the intervals (5.35) with $(2F^{\alpha}_{2,n-2})^{1/2}$ replacing $t^{\alpha/2}_{n-2}$ are good approximations to the exact intervals. Figures illustrating these ideas for the Working-Hotelling bands can be found in Miller (1981, pp. 118–119).

Similar comments hold for the Graybill-Bowden confidence band (5.27). When $\widehat{CV}(\hat{\beta})$ is quite small, the confidence interval is approximately

$$x(\mu) \in \frac{\mu - \bar{y}}{\hat{\beta}} \pm |m|^{\alpha}_{2,n-2} \frac{\hat{\sigma}}{|\hat{\beta}|} \left[\frac{1}{\sqrt{n}} + \frac{|\mu - \bar{y}|}{|\hat{\beta}| \sqrt{\sum_{i=1}^{n} (x_i - \bar{x})^2}} \right]. \tag{5.36}$$

Calibration problems often arise in the following context. Two ways of measuring the same quantity are available. One is very accurate, time-consuming, and possibly expensive; the other is more variable, easier to obtain, and usually cheaper. The one may be a direct measurement, and the other an indirect measurement. The laboratory develops a *standard line* by laboriously obtaining a series of paired values (x_i, y_i), $i = 1, \cdots, n$, where x_i is the direct or more accurate measurement and y_i is the indirect or more variable measurement. The regression line $\hat{\alpha} + \hat{\beta}x$ is estimated from these paired values. Further readings on unknown amounts of the substance are obtained by measuring just with the indirect or more variable procedure and converting the measured y to x by $\hat{x} = (y - \hat{\alpha})/\hat{\beta}$. Typically, a standard line is used to calibrate a number of additional measurements.

Usually only the point estimates $\hat{x} = (y - \hat{\alpha})/\hat{\beta}$ are of interest to the laboratory, but at times some idea of the variability in the method is desired. If the standard line is accurately determined with small $\widehat{CV}(\hat{\beta})$, then an approximate standard deviation for $\hat{x}$ is

$$\frac{\hat{\sigma}}{|\hat{\beta}|}\left[1 + \frac{1}{n} + \frac{(y - \bar{y})^2}{\hat{\beta}^2 \sum_{i=1}^{n}(x_i - \bar{x})^2}\right]^{1/2}. \tag{5.37}$$

Expression (5.37) is similar to the standard deviation factor appearing in (5.35). The difference is the extra "1" which enters the variance due to the variability in the single y measurement about its mean value $\mu(x)$. Note that the size of (5.37) will vary somewhat depending on whether the observed y is at the low or high ends of the range or in the middle.

An exact confidence interval for x_0 corresponding to a single additional observation y_0 is given by (5.32) with an additional "1" added inside the braces to include the variability in the y_0 measurement. Simultaneous confidence intervals corresponding to an arbitrary number of additional measurements exist (see Lieberman et al., 1967, and Scheffé, 1973) but seem to be rarely used.

The theory of calibration is extensive and has only been touched upon here. Alternative procedures, such as regression x on y (see Krutchkoff, 1967) or adopting a Bayesian approach (see Hoadley, 1970, and Hunter and Lamboy, 1981), exist in the literature. Historical perspective and references are given in the Hunter and Lamboy (1981) article and in the discussion articles that follow it, especially Rosenblatt and Spiegelman (1981).

5.1.2. One Sample: Zero Intercept

Not often, but every now and then, one knows that $\alpha = 0$ from a priori considerations about the experiment or a graphical plot that strongly indicates the data follow a ray emanating from the origin.

In such a situaion the reduced model is

$$E(y \mid x) = \mu(x) = \beta x \tag{5.38}$$

and

$$\text{Var}(y \mid x) = \sigma^2(x) \equiv \sigma^2. \tag{5.39}$$

Typically, the x_i and y_i values are all positive when this model is applied.

The maximum likelihood estimate of the slope is

$$\hat{\beta} = \frac{\sum_{i=1}^{n} x_i y_i}{\sum_{i=1}^{n} x_i^2}. \tag{5.40}$$

The bias-corrected maximum likelihood estimate of the variance is

$$\begin{aligned} \hat{\sigma}^2 &= \frac{1}{n-1} \sum_{i=1}^{n} (y_i - \hat{\beta} x_i)^2, \\ &= \frac{1}{n-1} \left[\sum_{i=1}^{n} y_i^2 - \frac{\left(\sum_{i=1}^{n} x_i y_i\right)^2}{\sum_{i=1}^{n} x_i^2} \right]. \end{aligned} \tag{5.41}$$

The estimator $\hat{\beta}$ has a normal distribution with mean β and variance

$$\text{Var}(\hat{\beta}) = \frac{\sigma^2}{\sum_{i=1}^{n} x_i^2}. \tag{5.42}$$

The variance estimate $\hat{\sigma}^2$ is distributed independently of $\hat{\beta}$ and has a scaled χ^2 distribution with $n - 1$ df, i.e.,

$$\frac{(n-1)\hat{\sigma}^2}{\sigma^2} \sim \chi^2_{n-1}. \tag{5.43}$$

For testing a null hypothesis $H_0 : \beta = \beta_0$, the ratio

$$\frac{\hat{\beta} - \beta_0}{\hat{\sigma} / \sqrt{\sum_{i=1}^{n} x_i^2}}, \tag{5.44}$$

which has a t distribution with $n-1$ df, provides P values. The associated $100(1-\alpha)\%$ confidence interval is

$$\beta \in \hat{\beta} \pm t_{n-1}^{\alpha/2} \frac{\hat{\sigma}}{\sqrt{\sum_{i=1}^{n} x_i^2}}. \tag{5.45}$$

Since the values of the mean function $\mu(x) = \beta x$ at different values of x are simply known scalar multiples of each other, there is no distinction between a confidence interval for a single x_0 and a confidence band for many x. From (5.45) it follows directly that

$$\mu(x) \in \hat{\beta} x \pm t_{n-1}^{\alpha/2} \frac{\hat{\sigma} x}{\sqrt{\sum_{i=1}^{n} x_i^2}} \tag{5.46}$$

with probability exactly $1-\alpha$ for any number of x.

In the calibration problem the inverse estimate for x is

$$\hat{x} = \frac{\mu}{\hat{\beta}}. \tag{5.47}$$

For a poorly determined $\hat{\beta}$ the biased reduced estimate

$$\frac{\mu}{\hat{\beta}} \left[1 - \frac{\hat{\sigma}^2}{\hat{\beta}^2 \sum_{i=1}^{n} x_i^2} \right] \tag{5.48}$$

might be more accurate. From (5.45) the confidence interval for $x(\mu)$ is simply

$$x(\mu) \in \frac{\mu}{\hat{\beta} \pm t_{n-1}^{\alpha/2} \hat{\sigma} / \sqrt{\sum_{i=1}^{n} x_i^2}}, \tag{5.49}$$

provided the values in the denominator have the same sign (usually positive). When the calibration involves a variable y, the inverse estimate from the standard line has the same form $\hat{x} = y/\hat{\beta}$ as (5.47), but the confidence intervals differ from (5.49). For an accurately determined standard line with small $\widehat{CV}(\hat{\beta}) = \hat{\sigma}/\hat{\beta}\left(\sum_{i=1}^{n} x_i^2\right)^{1/2}$, an approximate standard deviation for $\hat{x}$ is

$$\frac{\hat{\sigma}}{|\hat{\beta}|}\left[1+\frac{y^2}{\hat{\beta}^2\sum_{i=1}^n x_i^2}\right]^{1/2}. \tag{5.50}$$

5.1.3. Multisamples: General Intercepts

With I separate populations the full model is

$$y_{ij} = \alpha_i + \beta_i x_{ij} + e_{ij} \tag{5.51}$$

for $i = 1, \cdots, I$, $j = 1, \cdots, n_i$, where the e_{ij} are assumed to be independently, identically distributed as $N(0, \sigma^2)$.

In classical analysis of variance this model would be discussed under the heading *analysis of covariance*. It is a one-way classification in the population intercepts with a single *covariate* x_{ij}. Before the advent of large computers, specialized computational techniques were devised for analyzing experimental designs with single or multiple covariates. The computational techniques were based on the simple analyses of variance for the designs (viz., one-way ANOVA, etc.) relating the intercepts of the regression lines. With our current computers which can speedily spit out large multiple regression analyses, these specialized methods are no longer so relevant.

Usually the first major question to be addressed is "Are the slopes equal?" If $\beta_1 = \cdots = \beta_I$, then the family of I regression lines is conveniently restricted, and comparisons between regression lines greatly simplify. With unequal slopes, any bizarre collection of lines is possible with irregular criss-crossing like the game of "Pick Up Sticks." In general, whether the mean for one population is higher or lower than the mean for another depends on which values of the independent variable x are under consideration. With equal slopes, differences between populations are characterized solely by the differences in the intercepts $\alpha_1, \cdots, \alpha_I$.

A graphical plot of the data will frequently indicate whether the assumption of equal slopes is at all reasonable. A formal test of the null hypothesis $H_0 : \beta_1 = \cdots = \beta_I$ is carried out as follows. The estimate of the common slope β under the null hypothesis H_0 is

$$\hat{\beta} = \frac{\sum_{i=1}^{I} \sum_{j=1}^{n_i} (x_{ij} - \bar{x}_{i\cdot})(y_{ij} - \bar{y}_{i\cdot})}{\sum_{i=1}^{I} \sum_{j=1}^{n_i} (x_{ij} - \bar{x}_{i\cdot})^2}, \tag{5.52}$$

where $\bar{x}_{i\cdot} = (1/n_i) \sum_{j=1}^{n_i} x_{ij}$, etc. This is a weighted combination

$$\hat{\beta} = \frac{\sum_{i=1}^{I} w_i \hat{\beta}_i}{\sum_{i=1}^{I} w_i} \tag{5.53}$$

of the separate slope estimates

$$\hat{\beta}_i = \frac{\sum_{j=1}^{n_i} (x_{ij} - \bar{x}_{i\cdot})(y_{ij} - \bar{y}_{i\cdot})}{\sum_{j=1}^{n_i} (x_{ij} - \bar{x}_{i\cdot})^2}, \tag{5.54}$$

with weights

$$w_i = \sum_{j=1}^{n_i} (x_{ij} - \bar{x}_{i\cdot})^2. \tag{5.55}$$

Under H_0 the sum of the weighted squared differences

$$\sum_{i=1}^{I} w_i (\hat{\beta}_i - \hat{\beta})^2 \tag{5.56}$$

is distributed as σ^2 times a χ^2 variable with $I - 1$ df. When (5.56) is divided by $I - 1$ and the pooled estimate $\hat{\sigma}^2$, the ratio has an F distribution with $I - 1$ and $\sum_{i=1}^{I} (n_i - 2)$ df. Typically, one wants to accept the null hypothesis unless the P value calculated for the observed ratio is so small as to preclude this decision.

The pooled estimate $\hat{\sigma}^2$ referred to in the preceding paragraph is

$$\hat{\sigma}^2 = \frac{1}{N - 2I} \sum_{i=1}^{I} \sum_{j=1}^{n_i} (y_{ij} - \hat{\alpha}_i - \hat{\beta}_i x_{ij})^2, \tag{5.57}$$

where $N = \sum_{i=1}^{I} n_i$, $\hat{\beta}_i$ is given by (5.54), and

$$\hat{\alpha}_i = \bar{y}_{i\cdot} - \hat{\beta}_i \bar{x}_{i\cdot}\,. \tag{5.58}$$

The estimate $\hat{\sigma}^2$ is a weighted combination

$$\hat{\sigma}^2 = \frac{\sum_{i=1}^{I}(n_i - 2)\hat{\sigma}_i^2}{\sum_{i=1}^{I}(n_i - 2)} \tag{5.59}$$

of the separate error variance estimates

$$\hat{\sigma}_i^2 = \frac{1}{n_i - 2}\sum_{j=1}^{n_i}(y_i - \hat{\alpha}_i - \hat{\beta}_i x_{ij})^2 \tag{5.60}$$

for the different samples with weights equal to the respective degrees of freedom $n_i - 2$, $i = 1, \cdots, I$.

An alternative test of $H_0 : \beta_1 = \cdots = \beta_I$ is a Tukey-Kramer-type multiple comparisons procedure (see Section 3.1.2). This test would reject H_0 for large values of

$$\max_{i,i'} \left\{ \frac{|\hat{\beta}_i - \hat{\beta}_{i'}|}{\frac{\hat{\sigma}}{\sqrt{2}}\left[\frac{1}{\sum_{j=1}^{n_i}(x_{ij}-\bar{x}_{i\cdot})^2} + \frac{1}{\sum_{j=1}^{n_{i'}}(x_{i'j}-\bar{x}_{i'\cdot})^2}\right]^{1/2}} \right\}. \tag{5.61}$$

Under H_0 the distribution of (5.61) is approximately that of a studentized range of I variables with $N - 2I$ df for the error variance estimate. Tables of the studentized range appear in Harter (1960, 1969a), Miller (1981), Owen (1962), and Pearson and Hartley (1970).

If there were any reason to suspect monotone alternatives for the slopes in the event that the slopes were unequal, a statistic exploiting this information could be applied (see Section 3.1.3).

If the decision is made that the slopes are equal, then the analysis proceeds on the basis of the restricted model

$$y_{ij} = \alpha_i + \beta x_{ij} + e_{ij}. \tag{5.62}$$

The maximum likelihood estimator for β is given by (5.52). For $\alpha_1, \cdots, \alpha_I$ the MLE are

$$\hat{\alpha}_i = \bar{y}_{i\cdot} - \hat{\hat{\beta}}\bar{x}_{i\cdot}, \tag{5.63}$$

$i = 1, \cdots, I$, which differ from (5.58) because of the common slope estimate. The estimator for σ^2 becomes

$$\begin{aligned} \hat{\sigma}^2 &= \frac{1}{N-I-1}\sum_{j=1}^{I}\sum_{j=1}^{n_i}(y_{ij} - \hat{\alpha}_i - \hat{\hat{\beta}}x_{ij})^2, \qquad (5.64) \\ &= \frac{1}{N-I-1}\left[\sum_{i=1}^{I}\sum_{j=1}^{n_i} y_{ij}^2 - \sum_{i=1}^{I} n_i\bar{y}_{i\cdot}^2 - \hat{\hat{\beta}}^2\sum_{i=1}^{I}\sum_{j=1}^{n_i}(x_{ij} - \bar{x}_{i\cdot})^2\right]. \end{aligned}$$

The estimators $(\hat{\alpha}_1, \cdots, \hat{\alpha}_I)$ have a multivariate normal distribution with mean $(\alpha_1, \cdots, \alpha_I)$ and variances-covariances

$$\begin{aligned} \mathrm{Var}(\hat{\alpha}_i) &= \sigma^2\left[\frac{1}{n_i} + \frac{\bar{x}_{i\cdot}^2}{\sum_{i=1}^{I}\sum_{j=1}^{n_i}(x_{ij} - \bar{x}_{i\cdot})^2}\right], \\ \mathrm{Cov}(\hat{\alpha}_i, \hat{\alpha}_{i'}) &= \sigma^2\left[\frac{\bar{x}_{i\cdot}\ \bar{x}_{i'\cdot}}{\sum_{i=1}^{I}\sum_{j=1}^{n_i}(x_{ij} - \bar{x}_{i\cdot})^2}\right]. \end{aligned} \tag{5.65}$$

The variance estimator $\hat{\sigma}^2$ is independent of $(\hat{\alpha}_1, \cdots, \hat{\alpha}_I)$ and

$$\frac{(N-I-1)\hat{\sigma}^2}{\sigma^2} \sim \chi^2_{N-I-1}. \tag{5.66}$$

The classical ANOVA test of $H_0 : \alpha_1 = \cdots = \alpha_I$ compares the residual sum of squares under the model (5.62) with the corresponding residual sum of squares under H_0. Under H_0 the MLE for β changes to

$$\hat{\beta} = \frac{\sum_{i=1}^{I}\sum_{j=1}^{n_i}(x_{ij} - \bar{x}_{\cdot\cdot})(y_{ij} - \bar{y}_{\cdot\cdot})}{\sum_{i=1}^{I}\sum_{j=1}^{n_i}(x_{ij} - \bar{x}_{\cdot\cdot})^2}, \tag{5.67}$$

where $\bar{x}_{\cdot\cdot} = \sum_{i=1}^{I} \sum_{j=1}^{n_i} x_{ij}/N$, etc. The difference in the residual sums of squares is given by

$$SS(\boldsymbol{\alpha}; H_0) = \sum_{i=1}^{I} n_i \bar{y}_{i\cdot}^2 - N\bar{y}_{\cdot\cdot}^2 + \hat{\beta}^2 \sum_{i=1}^{I} \sum_{j=1}^{n_i} (x_{ij} - \bar{x}_{i\cdot})^2 \\ - \hat{\beta}^2 \sum_{i=1}^{I} \sum_{j=1}^{n_i} (x_{ij} - \bar{x}_{\cdot\cdot})^2. \tag{5.68}$$

Under H_0 the ratio

$$\frac{SS(\boldsymbol{\alpha}; H_0)}{(I-1)\hat{\sigma}^2} \tag{5.69}$$

has an F distribution with $I-1$ and $N-I-1$ df for numerator and denominator, respectively.

In many instances the independent variable means $\bar{x}_{i\cdot}$, $i = 1, \cdots, I$, are roughly equal and/or the squares of these means are small relative to $\sum_{i=1}^{I} \sum_{j=1}^{n_i} (x_{ij} - \bar{x}_{i\cdot})^2$. Either event ensures that

$$\begin{aligned} \operatorname{Var}(\hat{\alpha}_i - \hat{\alpha}_{i'}) &\cong \sigma^2 \left(\frac{1}{n_i} + \frac{1}{n_{i'}} \right), & i \neq i', \\ \operatorname{Cov}(\hat{\alpha}_i - \hat{\alpha}_{i'}, \hat{\alpha}_{i''} - \hat{\alpha}_{i'}) &\cong \frac{\sigma^2}{n_{i'}}, & i \neq i', i'', \end{aligned} \tag{5.70}$$

and the other covariances are all approximately zero. This covariance structure is identical to the one-way classification with unequal sample sizes, so the Tukey-Kramer method of multiple comparisons can be applied (see Section 3.1.2). With probability approximately $1 - \alpha$,

$$\alpha_i - \alpha_{i'} \in \hat{\alpha}_i - \hat{\alpha}_{i'} \pm q_{I,N-I-1}^{\alpha} \frac{\hat{\sigma}}{\sqrt{2}} \left[\frac{1}{n_i} + \frac{1}{n_{i'}} \right. \\ \left. + \frac{(\bar{x}_{i\cdot} - \bar{x}_{i'\cdot})^2}{\sum_{i=1}^{I} \sum_{j=1}^{n_i} (x_{ij} - \bar{x}_{i\cdot})^2} \right]^{1/2} \tag{5.71}$$

for all $i \neq i'$, where $q_{I,N-I-1}^{\alpha}$ is the upper α percentile for the studentized range of I variables with $N - I - 1$ df for the error variance estimate. The last ratio under the square root in (5.71) should be small relative to the two preceding terms for the coverage probability to be approximately correct.

For testing $H_0 : \alpha_1 = \cdots = \alpha_I$ against montone alternatives, one can use an appropriate contrast $\sum_{i=1}^{I} c_i \hat{\alpha}_i$ (see Section 3.1.3). The variance of $\sum_{i=1}^{I} c_i \hat{\alpha}_i$ can be derived from (5.65).

Although comparisons between populations with a common slope are usually characterized by differences between intercepts, in bioassay there is meaning in converting a difference in intercepts into a difference in x values. Specifically, for two populations the ratio

$$\Delta_{12} = \frac{\alpha_2 - \alpha_1}{\beta} \tag{5.72}$$

is the amount that must be added to an x value in population 1 to achieve the same effect as an x value in population 2, i.e.,

$$\mu_1(x + \Delta_{12}) = \mu_2(x), \tag{5.73}$$

where $\mu_i(x) = \alpha_i + \beta x$, $i = 1, 2$. If the x scale is, in fact, the logarithm of a drug dose, then $\rho_{12} = \exp(\Delta_{12})$ is called the *relative potency* of the two drugs and is the factor by which a dose level of drug 1 must be multiplied to produce the same effect as an identical amount of drug 2 (see Finney, 1978, pp. 79–80). The relative potency ρ_{12} can be greater or less than one depending on whether Δ_{12} is positive or negative.

Point and interval estimation for Δ_{12} is very similar to the calibration problem for just one population. The commonly used point estimate of Δ_{12} is

$$\hat{\Delta}_{12} = \frac{\hat{\hat{\alpha}}_2 - \hat{\hat{\alpha}}_1}{\hat{\hat{\beta}}}, \tag{5.74}$$

where $\hat{\hat{\beta}}$ is given by (5.52) and $\hat{\hat{\alpha}}_i$, $i = 1, 2$, by (5.63). If β cannot be accurately estimated, then nontrivial bias can creep into $\hat{\Delta}_{12}$ due to

$\hat{\hat{\beta}}$ being in the denominator, and the bias-adjusted estimator

$$\hat{\hat{\Delta}}_{12} = (\bar{x}_{1\cdot} - \bar{x}_{2\cdot}) + \left[\frac{\bar{y}_{2\cdot} - \bar{y}_{1\cdot}}{\hat{\hat{\beta}}}\right]\left[1 - \frac{\hat{\hat{\sigma}}^2}{\hat{\hat{\beta}}^2 \sum_{i=1}^{I}\sum_{j=1}^{n_i}(x_{ij} - \bar{x}_{i\cdot})^2}\right] \tag{5.75}$$

may offer an improved estimate. Fully exact Fieller intervals analogous to (5.32) are

$$\bar{x}_{1\cdot} - \bar{x}_{2\cdot} + \frac{(\bar{y}_{2\cdot} - \bar{y}_{1\cdot})}{\hat{\hat{\beta}}(1-\epsilon')} \pm \frac{(F^{\alpha}_{1,N-I-1})^{1/2}\hat{\hat{\sigma}}\left[\left(\frac{1}{n_1} + \frac{1}{n_2}\right)(1-\epsilon') + \frac{(\bar{y}_{2\cdot} - \bar{y}_{1\cdot})^2}{\sum_{i=1}^{I}\sum_{j=1}^{n_i}(x_{ij} - \bar{x}_{i\cdot})^2}\right]^{1/2}}{\hat{\hat{\beta}}(1-\epsilon')}, \tag{5.76}$$

where

$$\epsilon' = \frac{F^{\alpha}_{1,N-I-1}\hat{\hat{\sigma}}^2}{\hat{\hat{\beta}}^2 \sum_{i=1}^{I}\sum_{j=1}^{n_i}(x_{ij} - \bar{x}_{i\cdot})^2}. \tag{5.77}$$

Absurdities such as the confidence interval being the whole axis (in the case of imaginary roots) or two semi-infinite intervals [when $\hat{\Delta}_{12}$ lies outside the roots (5.76)] can occur just as for (5.32). For well-estimated β the factor ϵ' is small, and the intervals

$$\Delta_{12} \in \hat{\Delta}_{12} \pm t^{\alpha}_{N-I-1}\frac{\hat{\hat{\sigma}}}{|\hat{\hat{\beta}}|}\left[\frac{1}{n_1} + \frac{1}{n_2} + \frac{(\bar{y}_{2\cdot} - \bar{y}_{1\cdot})^2}{\sum_{i=1}^{I}\sum_{j=1}^{n_i}(x_{ij} - \bar{x}_{i\cdot})^2}\right]^{1/2} \tag{5.78}$$

give a good approximation to (5.76). The intervals (5.78) are always well defined, more intuitive, and easier to explain.

When there are more than just two populations, the preceding formulas apply for point and interval estimates of the log relative potency $\Delta_{ii'}$ between any two populations i and i'. With more than two populations, it may be the case that one population, say, $i = 1$, is a standard population against which all others are compared. This would limit the estimation to the $I - 1$ log relative potencies $\Delta_{12}, \cdots, \Delta_{1I}$. If it is important to have simultaneous confidence in all the intervals, the significance level α in (5.76) or (5.78) can be changed to α/K, where K is the number of relative potencies being considered [see(3.12)].

5.1.4. Multisamples: Zero Intercepts

In the model where the intercepts are known to be identically zero (i.e., $\alpha_i \equiv 0$), the individual slope estimates are

$$\hat{\beta}_i = \frac{\sum_{j=1}^{n_i} x_{ij} y_{ij}}{\sum_{j=1}^{n_i} x_{ij}^2}, \tag{5.79}$$

and the combined estimate of β under the hypothesis $H_0 : \beta_1 = \cdots = \beta_I$ is

$$\begin{aligned} \hat{\hat{\beta}} &= \frac{\sum_{i=1}^{I} w_i \hat{\beta}_i}{\sum_{i=1}^{I} w_i}, \\ &= \frac{\sum_{i=1}^{I} \sum_{j=1}^{n_i} x_{ij} y_{ij}}{\sum_{i=1}^{I} \sum_{j=1}^{n_i} x_{ij}^2}, \end{aligned} \tag{5.80}$$

where

$$w_i = \sum_{j=1}^{n_i} x_{ij}^2. \tag{5.81}$$

For different β_i the estimate of σ^2 is

$$\begin{aligned} \hat{\sigma}^2 &= \frac{1}{N - I} \sum_{i=1}^{I} \sum_{j=1}^{n_i} (y_{ij} - \hat{\beta}_i x_{ij})^2, \\ &= \frac{1}{N - I} \sum_{i=1}^{I} \left[\sum_{j=1}^{n_i} y_{ij}^2 - \frac{\left(\sum_{j=1}^{n_i} x_{ij} y_{ij}\right)^2}{\sum_{j=1}^{n_i} x_{ij}^2} \right]. \end{aligned} \tag{5.82}$$

The individual slope estimates $\hat{\beta}_i$ are independently normally distributed with means β_i and variances σ^2/w_i, $i = 1, \cdots, I$, respectively. The ratio $(N - I)\hat{\sigma}^2/\sigma^2$ has a χ^2 distribution with $N - I$ df and is independent of $\hat{\beta}_1, \cdots, \hat{\beta}_I$. When $H_0 : \beta_1 = \cdots = \beta_I$ is true, the combined estimate $\hat{\hat{\beta}}$ has a normal distribution with mean β and variance $\sigma^2/\sum_{i=1}^I w_i$.

The classical test of the null hypothesis $H_0 : \beta_1 = \cdots = \beta_I$ relies on the statistic

$$\frac{\sum_{i=1}^I w_i(\hat{\beta}_i - \hat{\hat{\beta}})^2}{(I-1)\hat{\sigma}^2}, \tag{5.83}$$

which has an F distribution with $I-1$ and $N-I$ df. A Tukey-Kramer type multiple comparisons procedure would compare

$$\max_{i,i'} \left\{ \frac{|\hat{\beta}_i - \hat{\beta}_{i'}|}{\frac{\hat{\sigma}}{\sqrt{2}}\left[\frac{1}{\sum_{j=1}^{n_i} x_{ij}^2} + \frac{1}{\sum_{j=1}^{n_{i'}} x_{i'j}^2}\right]^{1/2}} \right\} \tag{5.84}$$

with the precentage points of a studentized range of I variables with $N - I$ df for the error variance estimate (see Section 3.1.2). For monotone alternatives an appropriately selected (see Section 3.1.3) linear combination $\sum_{i=1}^I c_i\hat{\beta}_i$ with estimated variance $\hat{\sigma}^2 \sum_{i=1}^I c_i^2/w_i$ would yield a t statistic.

In a bioassay with two lines, the ratio $\rho_{12} = \beta_2/\beta_1$ gives the *relative potency* of the two preparations, namely, the factor by which a dose of preparation 1 must be multiplied to give the same rsponse as an identical dose of preparation 2. That is,

$$\mu_1(\rho_{12}x) = \mu_2(x). \tag{5.85}$$

The customary estimate of ρ_{12} is $\hat{\beta}_2/\hat{\beta}_1$, but one may want to make a bias correction in

$$\hat{\hat{\rho}}_{12} = \frac{\hat{\beta}_2}{\hat{\beta}_1}\left[1 - \frac{\hat{\sigma}^2}{\hat{\beta}_1^2 \sum_{j=1}^{n_1} x_{ij}^2}\right]. \tag{5.86}$$

A fully exact Fieller interval can be constructed by solving for the roots of

$$(\hat{\beta}_2 - \hat{\beta}_1 \rho_{12})^2 = F_{1,N-I}^2 \hat{\sigma}^2 \left(\frac{1}{\sum_{j=1}^{n_2} x_{2j}^2} + \frac{\rho_{12}^2}{\sum_{j=1}^{n_1} x_{1j}^2}\right), \tag{5.87}$$

but for a good assay the interval

$$\rho_{12} \in \hat{\rho}_{12} \pm t_{N-I}^{\alpha/2} \frac{\hat{\sigma}}{|\hat{\beta}_1|}\left[\frac{1}{\sum_{j=1}^{n_2} x_{2j}^2} + \frac{\hat{\beta}_2^2}{\sum_{j=1}^{n_1} x_{1j}^2}\right]^{1/2} \tag{5.88}$$

should suffice. For multiple relative potencies the significance level can be reduced to α/K, where K is the number of potencies being considered [see (3.12)].

In the parlance of bioassay this type of analysis with $\alpha_i \equiv 0$ is referred to as a *slope ratio assay*. Finney (1978, Chapter 7) studies the more general situation where $\alpha_i \equiv \alpha \neq 0$.

5.2. Nonlinearity.

God has not decreed that all regressions should be linear. Many are not. The mean regression function $\mu(x)$ might be quadratic $\alpha + \beta x + \gamma x^2$, exponential $\alpha e^{-\beta x}$, power αx^{β}, or something else.

If for reasons external to the data the form of the nonlinear regression is theoretically known, then one typically has two choices. The original scales for the variables can be maintained, and a nonlinear regression analysis can be applied. For an exposition of nonlinear regression analysis see Draper and Smith (1981, Chapter 10). In many instances one can also transform one or both variables so that the transformed relationship is linear. For example, with the exponential relation $y = \alpha e^{-\beta x}$, taking the logarithm of y produces a linear relation. Whether one uses a nonlinear analysis or a linear analysis after transformation depends on several factors, an important one being the availability of a good nonlinear regression computer routine. Often the linear analysis after transformation is quicker and easier even if the nonlinear routine is available. When both routes are equally open,the choice should depend on the appropriateness of the error structure. Do the errors seem more normal, homoscedastic, and free of outliers under the transformed or untransformed model? Even this criterion becomes blurred when one admits the possibility of weighted linear or nonlinear regression.

The remainder of the discussion in this section is focused on the situation where the model is not known for sure a priori.

5.2.1. Effect

The effect of your or the computer's blindly fitting a linear regression to data from a nonlinear model is that the fit of the line to the data will be poor. Point and interval estimates of α and β will be so much rubbish, and the estimates of μ and x in the prediction and calibration problems may be badly biased. How badly off you are depends on the range of x values. Over a narrow x range even an exponential or logarithmic function can be indiscernible from a linear function. However, over a broad range of x where the curvature of $\mu(x)$ is influential the miscalculation can be considerable.

In some data sets there is such substantial scatter in the y direction that it is a moot point as to whether the model is linear or nonlinear. A linear fit will do as well as anything for these data sets.

5.2.2. Detection

Detection of nonlinearity is usually by eye. A plot of the data with the estimated regression line $\hat{\mu}(x) = \hat{\alpha} + \hat{\beta}x$ drawn on it typically

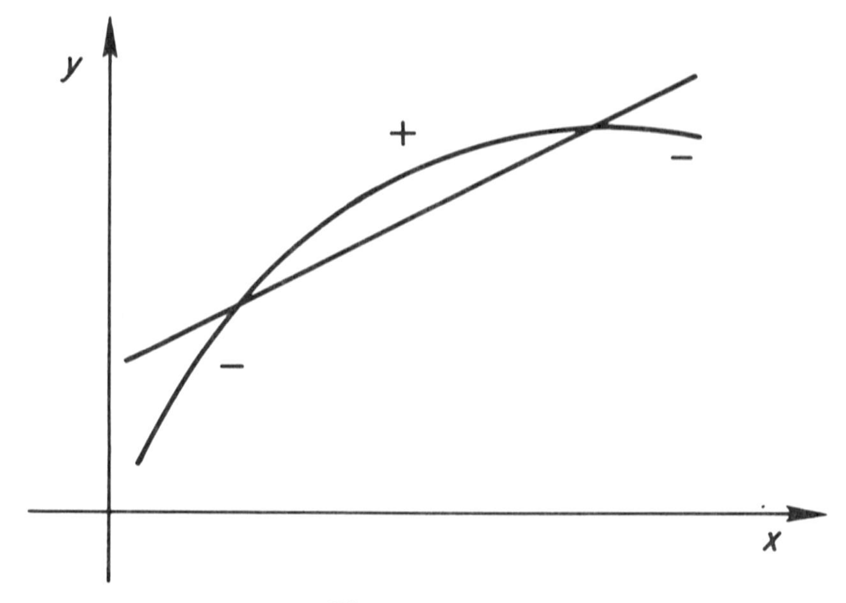

Figure 5.7

reveals whether the y values hover around the line over the whole range of x values. If the linear model holds, the y values should not be systematically above or below the estimated line for different regions of x values.

If a plot of the data is not readily available, the same sort of examination by eye can be performed on the residuals $r_i = y_i - \hat{\alpha} - \hat{\beta}x_i$, $i = 1, \cdots, n$. These should be ordered by their x values from small x to large x. If these residuals exhibit long runs of positive values alternating with long runs of negative values, there is evidence of nonlinearity. For example, if you try to fit a line to data from a concave quadratic regression function, the residuals tend to be negative for low x, positive in the middle, and negative for high x. This is illustrated schematically in Figure 5.7. The opposite pattern of $+ - +$ signs holds for convex quadratic functions.

Plotting the pairs (x_i, r_i), $i = 1, \cdots, n$, of independent variable values and residuals is a very effective method for spotting nonlinearity. If the model is correct, the residuals should jump randomly above and below the x-axis and not exhibit any discernible pattern.

This sort of human inspection is not possible when a decision on linearity vs. nonlinearity needs to be made automatically by a computer. If a large number of regression lines need to be estimated routinely, one may want the computer to compute each estimated line and flag those for which there is evidence of nonlinearity. One could try to mimic the eye inspection internally in the computer, but an alternate approach is simpler and probably better. The approach is to embed the linear model in a larger model with an additional parameter (or parameters) which for nonzero values produces nonlinearity in the model. For example, $\alpha + \beta x$ is a special case of $\alpha + \beta x + \gamma x^2$, and increasing the value of $|\gamma|$ induces increasing curvature into the model. The particular choice of a larger model usually reflects interplay between computational simplicity and the type of nonlinear departures one is anticipating. Once the larger model is selected, the data are tested for nonlinearity by testing whether the added curvature parameter is zero. Execution of this leads into multiple or nonlinear regression.

This latter numerical approach can be applied as well even when visual inspection is possible. It may be desirable in equivocal cases, but often a plot of the data can settle the issue without further calculation.

5.2.3. Correction

The correction for nonlinearity is to change the model. There are several ways to accomplish this.

One can attempt to retain the simple straight line analysis by transforming either the x or y variable. Which variable is selected depends on what transformation will linearize the data. For instance, with slowly increasing concave shaped positive data the log transformation of the x-axis from the model $y = \alpha + \beta x$ to $y = \alpha + \beta z$ where $z = \log x$ may produce linear data. On the other hand, for positive data which decays away as x increases the change from $y = \alpha + \beta x$ to $z = \alpha + \beta x$ where $z = \log y$ may yield a linear fit.

There are various graph papers that will assist you in searching for or checking on a linearizing transformation. Log-linear paper allows one to examine the effects of transforming either x or y by logarithms. Log-log paper gives a check of the model $y = \alpha x^{\beta}$. For bounded response variables like percentages or proportions, sigmodial regression functions are common, and probit and logistic papers with linear or log scales are useful in checking on the model.

It is more common to transform the y variable than the x variable. The logarithmic transformation $z = \log y$ is frequently used, and another common one is the reciprocal transformation $z = 1/y$. Rather than using a hit-or-miss search for the appropriate transformation one can use the systematic, analytical approach of Box and Cox (1964). For positive data (i.e., $y > 0$) they consider the family of power transformations

$$z = y^{(\lambda)} = \begin{cases} \frac{y^{\lambda}-1}{\lambda}, & \lambda \neq 0, \\ \log y, & \lambda = 0, \end{cases} \tag{5.89}$$

in conjunction with the linear model and suggest maximizing the normal likelihood or adopting a Bayesian analysis. Specifically, except for constants the log likelihood of the observations maximized over α and β is

$$\begin{aligned} \log L_{\max}(\lambda) = -\frac{n}{2} \log \left[\sum_{i=1}^{n} (y_i^{(\lambda)} - \hat{\alpha}^{(\lambda)} - \hat{\beta}^{(\lambda)} x_i)^2 \right] \\ + (\lambda - 1) \sum_{i=1}^{n} \log y_i, \end{aligned} \tag{5.90}$$

where $\hat{\alpha}^{(\lambda)}$ and $\hat{\beta}^{(\lambda)}$ are the usual intercept and slope estimators (5.6) applied to $y_i^{(\lambda)}$, $i = 1, \cdots, n$. A numerical search routine will reveal the maximizing value of λ or a close approximation thereof.

As stated for (5.89), the power transformation family is applicable only for positive observations. For data that include some negative values it may be possible to add a constant c to make all the data positive before transformation as in (1.22).

For greater detail on the Box-Cox method of locating a linearizing transformation the reader should consult their original article. Andrews (1971b), Atkinson (1973), and Carroll (1980) consider tests associated with the power parameter. The reader should be cautioned that the Box-Cox estimates are nonrobust. This is discussed in Section 5.3, along with the controversy over the appropriate variability estimates for $\hat{\alpha}^{(\lambda)}$ and $\hat{\beta}^{(\lambda)}$.

After the data have been transformed to linearity, the questions of normality, homoscedasticity, and independence of the errors need to be addressed. If the errors were normally distributed with equal variances for the original data, they may not be after the transformation. It is hoped that the reverse will be true. In many instances the data look more normal and homoscedastic after the transformation than before. Achieving the correct model takes precedence over compliance with assumptions about the error structure.

In some problems it is not possible to find a satisfactory linearizing transformation. The correct model may really be quadratic or a mixture of exponentials or some complicated function. This is by no means the end of the world. It simply means you need to penetrate beyond the scope of this book into the realms of multiple regression and nonlinear regression. For guidance see Draper and Smith (1981). It also means you will probably need a large computer, particularly for nonlinear regression.

5.3. Nonnormality.

5.3.1. Effect

The effects on the intercept and slope estimates and their distributions from sampling underlying distributions with nonzero kurtoses are relatively minimal. The impact of distributions with tails that are somewhat longer or shorter than the normal or are skewed is similar to the one sample problem; see Section 1.2.1.

The effects on the sample correlation coefficient are more pronounced. The null distribution of r [namely, that (5.22) has a t distribution] is relatively undisturbed by sampling from nonnormal distributions with nonzero skewness and/or kurtosis. Thus tests of $\rho = 0$ are relatively robust. However, the validity of the asymptotic variance for (5.23) being equal to $1/(n-3)$ depends crucially on the assumption of normality and can be quite different for nonnormal distributions. This makes confidence interval construction for ρ sensitive to the assumption of normality. For a quantitative assessment of these effects and earlier references by E. S. Pearson and others, see Duncan and Layard (1973).

Outliers are a disaster story. They can be really troublesome. It does not matter whether they are generated by a heavy-tailed distribution such as the Cauchy or by a contaminated normal distribution. If there are points that lie at a distance from the body of the data, they can exert an undue influence on the estimates. This is particularly true for the slope estimate $\hat{\beta}$. The relative position of the x value(s) associated with the outlier(s) in relation to the other observed x values plays a crucial role. Outliers with x near the ends of the x range unduly increase or decrease $\hat{\beta}$. On the other hand, outliers with x near the middle of the range have little impact on $\hat{\beta}$. However, these can still affect $\hat{\alpha}$.

These qualitative assertions about outliers can be quantified through examining the effect of deleting an observation (x_i, y_i). For a multiple linear regression model $\mathbf{y} = \mathbf{X}\boldsymbol{\beta} + \mathbf{e}$ with full rank, the estimator $\hat{\beta}_{-i}$ with the ith observation deleted is related to the full estimator $\hat{\boldsymbol{\beta}} = (\mathbf{X}^T\mathbf{X})^{-1}\mathbf{X}^T\mathbf{y}$ by

$$\hat{\boldsymbol{\beta}} - \hat{\boldsymbol{\beta}}_{-i} = \frac{r_i}{1 - h_{ii}}(\mathbf{X}^T\mathbf{X})^{-1}\mathbf{x}_i^T, \tag{5.91}$$

where $\mathbf{X}$ is the $n \times p$ design matrix with rows $\mathbf{x}_k$, $k = 1, \cdots, n$, of independent variables (usually $x_{k1} \equiv 1$), $h_{ii} = \mathbf{x}_i(\mathbf{X}^T\mathbf{X})^{-1}\mathbf{x}_i^T$ is the ith diagonal element of the hat matrix $\mathbf{H} = \mathbf{X}(\mathbf{X}^T\mathbf{X})^{-1}\mathbf{X}^T$, and $r_i = y_i - \mathbf{x}_i\hat{\boldsymbol{\beta}}$ is the ith residual (see Miller, 1974b, Lemma 3.2). In the case of linear regression (5.91) reduces to

$$\begin{aligned} \hat{\alpha} - \hat{\alpha}_{-i} &= \frac{r_i}{1 - h_{ii}}\left(\frac{1}{n} - \frac{(x_i - \bar{x})\bar{x}}{S_{xx}}\right), \\ \hat{\beta} - \hat{\beta}_{-i} &= \frac{r_i}{1 - h_{ii}}\left(\frac{x_i - \bar{x}}{S_{xx}}\right), \end{aligned} \tag{5.92}$$

where

$$\begin{aligned} r_i &= y_i - \hat{\alpha} - \hat{\beta}x_i, \\ S_{xx} &= \sum_{k=1}^{n}(x_k - \bar{x})^2, \\ h_{ii} &= \frac{1}{n} + \frac{(x_i - \bar{x})^2}{S_{xx}}. \end{aligned} \tag{5.93}$$

The change from $\hat{\beta}_{-i}$ to $\hat{\beta}$ in (5.92) is easily interpretable. The larger the absolute value of the residual r_i is, the greater the change will be, but the amount of change is influenced by the position of x_i relative to $\bar{x}$. This enters both through $(x_i - \bar{x})/S_{xx}$ and through $1 - h_{ii}$. The larger $|x_i - \bar{x}|$ is, the greater the change will be, with no change whatsoever when $x_i = \bar{x}$.

Cook (1977, 1979) proposed as an overall criterion for judging the influence of the ith observation the ratio

$$D_i = \frac{(\hat{\boldsymbol{\beta}} - \hat{\boldsymbol{\beta}}_{-i})^T(\mathbf{X}^T\mathbf{X})(\hat{\boldsymbol{\beta}} - \hat{\boldsymbol{\beta}}_{-i})}{p\hat{\sigma}^2}, \qquad (5.94)$$
$$= \frac{r_i^2 h_{ii}}{(1 - h_{ii})^2 p\hat{\sigma}^2},$$

where $\hat{\sigma}^2$ equals $\mathbf{y}^T(\mathbf{I} - \mathbf{H})\mathbf{y}/(n - p)$ in general and (5.8) in the case of simple linear regression with $p = 2$. Clearly, the size of D_i is affected by the magnitudes both of the residual r_i and of $h_{ii}/(1 - h_{ii})$, which measures how centrally located the x_i value is. Hoaglin and Welsch (1978) have suggested separately examining the h_{ii} for high values to identify points of high leverage on the estimates and the r_i to determine whether leverage has been applied.* Others, notably Box and Draper (1975), Davies and Hutton (1975), and Huber (1973, 1975), have also contended that large h_{ii} identify points of sensitivity in the design.

Andrews (1971b) was the first to sound the alarm that the Box-Cox procedure for selecting a transformation is sensitive to outliers. The estimates of λ and (α, β) are unstable under small perturbations of the data. Andrews (1971b) and Carroll (1980) proposed more robust tests.

Bickel and Doksum (1981) established by asymptotics and simulations that $\hat{\lambda}$ and $(\hat{\alpha}(\hat{\lambda}), \hat{\beta}(\hat{\lambda}))$ [i.e., $(\hat{\alpha}, \hat{\beta})$ computed from $y^{(\hat{\lambda})}$] are highly correlated and $(\hat{\alpha}(\hat{\lambda}), \hat{\beta}(\hat{\lambda}))$ has a substantial extra variance component due to λ being estimated. This has raised a controversy as to whether one should make inferences on the regression parameters unconditionally as in the Bickel-Doksum theory or whether one should operate conditionally given the value of $\hat{\lambda}$ (see Box and Cox, 1982). The latter is the procedure if one chooses a linearizing trans-

---

* Note that $0 \leq h_{ii} \leq 1$ and $\sum_{i=1}^{n} h_{ii} = p$; thus p/n is an average value for h_{ii}.

formation by the hit-or-miss search method. Carroll and Ruppert (1981b) have shown that in the prediction problem there is only a small increase in mean squared error due to not knowing λ in estimating the conditional median of y on the original scale given x_0. Also, Doksum and Wong (1983) have established that the usual tests of hypotheses behave as though λ were known in terms of level and power.

5.3.2. Detection

Detection of outliers can usually be accomplished by eye from a plot of the data. Also, the impact of an outlier on the slope estimate can be judged by noting how far away the x_i associated with the offending y_i lies from $\bar{x}$. Quantitative assessment of a potential outlier and its impact is embodied in the residuals $r_i = y_i - \hat{\alpha} - \hat{\beta}x_i$ and the hat matrix diagonal values h_{ii}, $i = 1, \cdots, n$ [see (5.93)]. The r_i and h_{ii} are particularly useful when visual inspection is not possible. Formal tests of significance for outliers in regression are considered by Andrews (1971a). The work of Andrews and others is fully discussed in Barnett and Lewis (1978, Section 7.3).

Detection of distributions more or less kurtotic than the normal is not so important because the effects on the regression estimates and tests are minimal. However, one can make a probit plot of the residuals r_i, $i = 1, \cdots, n$. (See Section 1.2.2 if probit plotting is unfamiliar.) The r_i are correlated due to the subtraction of the estimated regression line values, but the empirical cdf of the residuals is a consistent estimate of the underlying distribution function (see Duan, 1981); Pierce and Kopecky (1979) and Pierce and Gray (1982) consider goodness-of-fit tests in the regression setting.

5.3.3. Correction

For the correction of nonnormality there are alternative nonparametric regression procedures based on the median rather than the mean.

G. W. Brown and Mood's (1951) regression coefficient estimators are obtained by dividing the x_i values into two groups at their median m_x and then solving the equations

$$\begin{aligned} \underset{x_i<m_x}{\text{median}}\{y_i - a - bx_i\} &= \underset{x_i>m_x}{\text{median}}\{y_i - a - bx_i\}, \\ \underset{1\le i\le n}{\text{median}}\{y_i - a - bx_i\} &= 0, \end{aligned} \tag{5.95}$$

for $a = \hat{\alpha}$, $b = \hat{\beta}$. For distinct x_i Theil (1950) introduced the slope estimator

$$\hat{\beta} = \underset{x_i<x_j}{\text{median}}\left\{\frac{y_j - y_i}{x_j - x_i}\right\}; \tag{5.96}$$

Sen (1968) generalized the Theil estimator to nondistinct x_i. The Theil-Sen estimator is described in Hollander and Wolfe (1973, Chapter 9). Andrews (1974) proposed a robust estimator based on medians, and A. Siegel (1982) has a robust repeated medians estimator. However, these median-based estimators are seldom used in practice. In estimating regression coefficients, consumers usually do not worry about a lack of normality – with the exception of concern about outliers.

Outliers need to be reckoned with because of their possible substantial impact on the regression coefficient values. Most practitioners fit the least squares line, examine the residuals, trim any observations that appear to be outlying and influential, and refit the least squares line to the observations left after trimming. With very few trimmed observations, the variance estimates are typically computed as though the remaining untrimmed observations constituted the whole sample.

In an important paper Ruppert and Carroll (1980) have tried to formalize this process and study it. Their results are disturbing. This procedure is inefficient for normal or near normal distributions and also for very heavily contaminated distributions. In the latter case the outliers tend to mask themselves by substantially distorting the initial least squares estimate of the regression line. In addition, the asymptotic variance is not analogous to a trimmed mean because of a component that depends upon the estimator used to fit the line initially. This component is impossible to estimate without enormous samples because it requires density estimation, and its effect is nonnegligible. This leaves the aforementioned trimming procedure in a very unsatisfactory state unless the amount of trimming is very minimal.

Ruppert and Carroll (1980) have identified an initial estimator for the regression line that gives the trimmed sample regression estimator an asymptotic variance analogous to a trimmed mean. Unfortunately, this initial estimator requires specialized computation involving "regression quantiles" as defined by Koenker and Bassett (1978).

A variety of other robust regression estimators have been proposed and championed by different investigators. Bickel (1973) considered a class of L-estimators (see Section 1.2.3, "Robust Estimation," for terminology.) Various M-estimators for regression coefficients corresponding to different ψ functions have appeared in the literature. Huber (1977, 1981) gives a general discussion, and Gross (1977) studies the bisquare estimator in considerable detail. Also, R-estimators are well represented. For references on R-estimators see Bickel (1973) and Jurečková (1977). Most of these L, M, and R-estimators are computationally cumbersome. Since packaged programs are not commonly available, they are seldom used in practice.

As noted at the beginning of this section, the nonnull distribution of the sample correlation coefficient is sensitive to departures from normality. If one is wedded to the product moment correlation r given by (5.9), then far more robust confidence intervals can be constructed by jackknifing or bootstrapping the transformed correlation $\tanh^{-1} r$. For descriptions of these procedures and their assessment see Duncan and Layard (1973) and Efron (1981). It should be mentioned that jackknifing is not resistant to outliers. Trimming is best to remove their effects (see Hinkley, 1978, and Hinkley and Wang, 1980). Devlin et al. (1975) consider the general problem of robust estimation of correlation coefficients.

An alternative estimator used with some frequency for nonnormal looking data is *Kendall's coefficient* τ. This is a nonparametric measure of the degree of association between x and y used in lieu of the correlation coefficient. In spirit it is related to the Mann-Whitney form of the two sample Wilcoxon rank statistic.

Define

$$T(x_i, x_j; y_i, y_j) = \begin{cases} 1 & \text{if } (x_i - x_j)(y_i - y_j) > 0, \\ 0 & \text{if } (x_i - x_j)(y_i - y_j) < 0. \end{cases} \tag{5.97}$$

The function T is an indicator function that scores 1 for concordant pairs in which $x_i - x_j$ and $y_i - y_j$ both have the same sign and scores 0 for discordant pairs. Let

$$T = \frac{1}{\binom{n}{2}} \sum_{i=1}^{n-1} \sum_{j=i+1}^{n} T(x_i, x_j; y_i, y_j), \tag{5.98}$$

which is an estimate of

$$p = P\{(x_1 - x_2)(y_1 - y_2) > 0\}. \tag{5.99}$$

The statistic

$$\hat{\tau} = 2\left(T - \frac{1}{2}\right) \tag{5.100}$$

estimates Kendall's τ coefficient

$$\tau = p - (1 - p), \tag{5.101}$$

which varies from -1 to $+1$ like the correlation coefficient and measures the association between x and y. For the bivariate normal distribution τ is related to ρ by

$$\tau = \frac{2}{\pi} \sin^{-1} \rho. \tag{5.102}$$

Hollander and Wolfe (1973, Table A.21) give the upper tail of the cdf for $K = \binom{n}{2}\hat{\tau}$ under the null hypothesis of no association for $n = 4(1)40$. Owen (1962) gives a smaller table [viz., $n = 2(1)12$] for T. Asymptotically, $\hat{\tau}$ has a normal distribution with mean and variance

$$E(\hat{\tau}) = 0,$$
$$\text{Var}(\hat{\tau}) = \frac{2(2n+5)}{9n(n-1)}, \tag{5.103}$$

under H_0. P values for testing no association between x and y are readily obtained from the small sample tables or the large sample normal approximation. A confidence interval for τ can be computed as well; for details see Hollander and Wolfe (1973, Chapter 8).

When ties are present for either the x or y observations or both, the score function should assign the value 1/2 when $(x_i - x_j)(y_i - y_j) = 0$.* For a small number of ties the effect on the null distribution is minimal. However, for larger numbers of ties the null variance is smaller than (5.103). For a corrected variance see Hollander and Wolfe (1973, p. 187).

There are other nonparametric measures of association. A prominent one is Spearman's rank correlation coefficient in which the observations are replaced by their ranks and the usual Pearson product moment correlation (5.9) is then computed. For details on Spearman's coefficient and references on the other nonparametric measures available, consult Holland and Wolfe (1973) or most any standard nonparametric textbook.

---

* Some texts recommend changing the denominator in T as well; see Gibbons (1971).

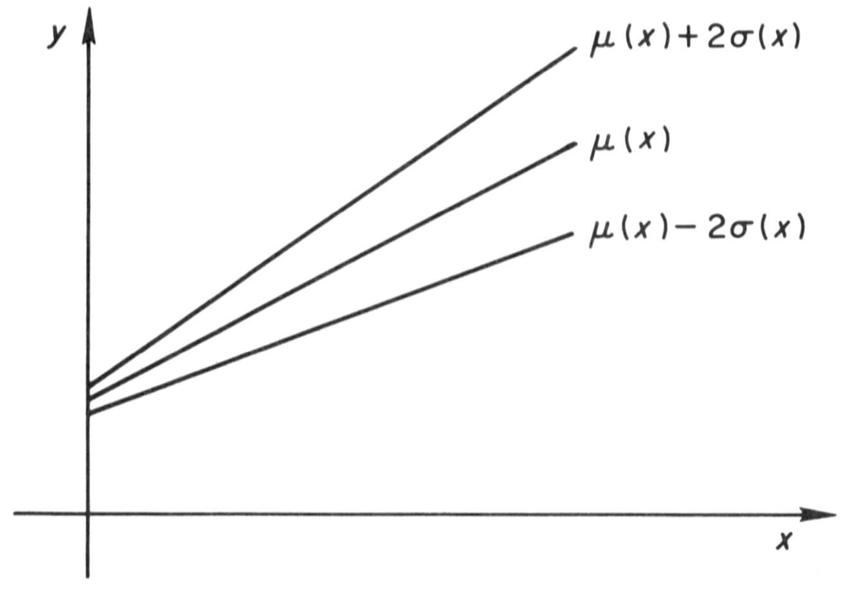

Figure 5.8

When one or both variables are binary or dichotomized, there are special correlation coefficients defined to cover these situations; see Walker and Lev (1953, Chapter 11) and Bishop et al. (1975, Section 11.2).

5.4. Unequal Variances.

The standard linear regression analysis is based on the assumption that the errors e_i in (5.1) all have the same variance σ^2. This, of course, need not be the case. Although any sort of heteroscedasticity is possible, it is likely that if there is a departure from equality it will be monotonically related to the regression mean value. Specifically, for data with positive x and y, the variance of y may tend to increase as the mean of y increases. This phenomonen is illustrated in Figure 5.8 where the shaded region indicates plus and minus two standard deviations.

The situation where the error variance is related to the mean regression has been modeled in the literature by

$$\mathrm{Var}(e_i) = \sigma_i^2 = \sigma^2 V(\mu(x_i), \lambda) \tag{5.104}$$

where, for example,

$$V(\mu(x_i), \lambda) = |\mu(x_i)|^\lambda \tag{5.105}$$

is a family of possible choices for V (see Box and Hill, 1974). Important special cases of (5.105) are $\lambda = 1$ and 2.

5.4.1. Effect

The effects of nonhomogeneity of variance are usually not dramatic unless the disproportionality between the variances is particularly severe. Even with unequal variances the usual least squares estimators (5.6) are unbiased, and under a variety of mild assumptions they are consistent. They no longer have any optimality properties, but most practitioners do not lose too much sleep over this. In multiple regression there are rare occasions in which the standard least squares estimator coincides with the best linear unbiased estimator (see Zyskind, 1967; Watson, 1967; Kruskal, 1968).

For normal errors the intercept and slope estimators are normally distributed, and for nonnormal errors they can be asymptotically normally distributed. The troublesome aspect is that the estimates of their variances are screwed up. The mean squares for error $\hat{\sigma}^2$ given by (5.8) estimates a mixture of different variances and the formulas (5.10) no longer apply. For example, the actual variance of $\hat{\beta}$ is

$$\frac{\sum_{i=1}^n (x_i - \bar{x})^2 \sigma_i^2}{\left[\sum_{i=1}^n (x_i - \bar{x})\right]^2}, \tag{5.106}$$

and this is estimated by $\hat{\sigma}^2 / \sum_{i=1}^n (x_i - \bar{x})^2$. However, if the weighted and unweighted averages of the variances are approximately equal, i.e.,

$$\frac{\sum_{i=1}^{n}(x_i - \bar{x})^2\sigma_i^2}{\sum_{i=1}^{n}(x_i - \bar{x})^2} \cong \frac{1}{n}\sum_{i=1}^{n}\sigma_i^2, \tag{5.107}$$

then the usual estimate for the variance of $\hat{\beta}$ should not be grossly inaccurate. This should be the case if the x_i are roughly symmetrically distributed about $\bar{x}$.

For a regression where the intercept α is known to be zero, the effect on the variance of $\hat{\beta}$ from a fan-shaped error structure such as Figure 5.8 is considerably worse. The actual variance of $\hat{\beta}$ is

$$\frac{\sum_{i=1}^{n} x_i^2\sigma_i^2}{\left[\sum_{i=1}^{n} x_i^2\right]^2}, \tag{5.108}$$

whereas the usual estimate of its variance is approixmately estimating

$$\frac{\frac{1}{n}\sum_{i=1}^{n}\sigma_i^2}{\sum_{i=1}^{n} x_i^2}. \tag{5.109}$$

Since large σ_i correspond to large x_i, the weighted average $\sum_{i=1}^{n} x_i^2\sigma^2 / \sum_{i=1}^{n} x_i^2$ definitely exceeds the unweighted average $\sum_{i=1}^{n}\sigma_i^2/n$, so (5.108) is larger than (5.109), possibly considerably larger. Thus the usual estimator for the variance of $\hat{\beta}$ may be badly underestimating the true variability.

5.4.2. Detection

Detection of heteroscedasticity is often relatively easy through a plot of the data or examination of the residuals. The fan-shaped behavior depicted in Figure 5.8 is readily detectable from a scattergram of the data. Other sorts of nonhomogeneity of variances can usually be spotted as well. When a graphical display is not available, or even when it is, the residuals can be examined. If they are ordered according to increasing values of x, the fan-shaped errors are revealed by the increasing size of the residuals. Other forms of inequality of variances lead to systematic changes in the sizes of the residuals.

There are formal tests for the homogeneity of the error variances based on analysis of the residuals. I do not feel a need to use them because if the disparity in the variances is not blatantly obvious from a scattergram or the residuals, it is not worth trying to correct. However, for readers who are less laid back about the problem and would like a significance test the following references should be valuable: Anscombe (1961), Bickel (1978), Carroll and Ruppert (1981a), and Cook and Weisberg (1983). Anscombe assumes normally distributed errors. Bickel does as well, but he also considers more robust tests as do Carroll and Ruppert.

5.4.3. Correction

If correction for heteroscedasticity is the prudent course of analysis, one might be extraordinarily lucky and find a transformation that both stabilizes the variances and creates a more linear model. However, in most instances one has to resort to a weighted least squares analysis.

In a *weighted least squares analysis* the sum of squares to be minimized is

$$\sum_{i=1}^{n} w_i(y_i - \alpha - \beta x_i)^2 \tag{5.110}$$

where $w_i = 1/\sigma_i^2$. If God or the Devil were willing to tell us the values for w_i, $i = 1, \cdots, n$, the solution would be

$$\begin{aligned} \hat{\alpha}_w &= \bar{y}_w - \hat{\beta}_w \bar{x}_w, \\ \hat{\beta}_w &= \frac{\sum_{i=1}^n w_i(x_i - \bar{x}_w)(y_i - \bar{y}_w)}{\sum_{i=1}^n w_i(x_i - \bar{x}_w)^2}, \end{aligned} \tag{5.111}$$

where

$$\bar{x}_w = \frac{\sum_{i=1}^n w_i x_i}{\sum_{i=1}^n w_i}, \quad \bar{y}_w = \frac{\sum_{i=1}^n w_i y_i}{\sum_{i=1}^n w_i}. \tag{5.112}$$

Since most of us cannot get help from above or below, we are faced with having to estimate the unknown weights.

One approach is to select a sensible system of weights involving the unknown regression parameters. For example, with positive data the variance at x may be roughly proportional to the mean of x, so $w_i = (\alpha + \beta x_i)^{-1}$. For constant coefficient of variation $w_i = (\alpha + \beta x_i)^{-2}$. Both are special cases of (5.105). After selection of an appropriate weight structure, minimization of (5.110) can be achieved through an iterative process. Initial estimates $\hat{\alpha}_0$, $\hat{\beta}_0$ (e.g., unweighted least squares estimates) are substituted for α, β in w_i, and for the first iteration the estimates $\hat{\alpha}_1$, $\hat{\beta}_1$ are calculated from (5.111) and (5.112). The one step estimates $\hat{\alpha}_1$, $\hat{\beta}_1$ are then used as the initial estimates in the weights, and the process is repeated. Convergence of the sequence of estimates $\hat{\alpha}_k$, $\hat{\beta}_k$, $k = 1, 2, \cdots$, to $\hat{\alpha}_{\hat{w}}$, $\hat{\beta}_{\hat{w}}$ is not a problem for smooth weight functions, but local, rather than global, minima can be troublesome.

Amemiya (1973) evalutes the performance of the one step estimates $\hat{\alpha}_1$, $\hat{\beta}_1$, where the aforementioned iterative procedure is stopped after the first step. Bement and Williams (1969), Jacquez et al. (1968), and Fuller and Rao (1978) also study one step estimators for the related problem where there are multiple observations at each distinct x value. When there are repeated y values at each different x, sample variance estimates (e.g., s_i^2) can be substituted for the unknown σ_i^2.

Under mild conditions on the σ_i^2 and the x_i, the estimators $\hat{\alpha}_{\hat{w}}$, $\hat{\beta}_{\hat{w}}$ are asymptotically normally distributed as $n \to \infty$. Their variances and covariances are estimated by

$$
\begin{aligned}
\widehat{\text{Var}}(\hat{\alpha}_{\hat{w}}) &= \hat{\sigma}_{\hat{w}}^2 \left[\frac{1}{\sum_{i=1}^n \hat{w}_i} + \frac{\bar{x}_{\hat{w}}^2}{\sum_{i=1}^n \hat{w}_i (x_i - \bar{x}_{\hat{w}})^2} \right], \\
\widehat{\text{Cov}}(\hat{\alpha}_{\hat{w}}, \hat{\beta}_{\hat{w}}) &= \hat{\sigma}_{\hat{w}}^2 \left[\frac{-\bar{x}_{\hat{w}}}{\sum_{i=1}^n \hat{w}_i (x_i - \bar{x}_{\hat{w}})^2} \right], \qquad (5.113) \\
\widehat{\text{Var}}(\hat{\beta}_{\hat{w}}) &= \hat{\sigma}_{\hat{w}}^2 \left[\frac{1}{\sum_{i=1}^n \hat{w}_i (x_i - \bar{x}_{\hat{w}})^2} \right],
\end{aligned}
$$

where estimates $\hat{w}_i$ of the weights w_i are used in the calculation and

$$\hat{\sigma}^2_{\hat{w}} = \frac{1}{n-2}\sum_{i=1}^{n} \hat{w}_i(y_i - \hat{\alpha}_{\hat{w}} - \hat{\beta}_{\hat{w}} x_i)^2. \qquad (5.114)$$

The rationale for using $n-2$ in the denominator of (5.114) has been totally lost; thus n might just as well be used instead. Also, the basis for using t distribution critical points has disappeared; normal critical values are justified by the asymptotics.

Instead of using weights involving the unknown parameters, empirical weights can be substituted. For example, if the variances are considered to be proportional to the square of the means, rather than using $w_i = (\alpha + \beta x_i)^{-2}$ in the iterative process, one can substitute $w_i = y_i^{-2}$. If there are values of y_i close to zero, the estimation may be improved by adding a small positive constant to the observed y_i in the weights [i.e., $w_i = (c + y_i)^{-2}$]. The use of empirical weights has the big advantage of eliminating the necessity for iteration.

My first preference is to use an unweighted analysis unless absolutely forced by the data to abandon it. In my experience unweighted estimates tend to be more stable than weighted estimates in the sense that small perturbations in the data do not produce much change in the estimates. However, if a weighted analysis is the order of the day, I would be more likely to use empirical weights than go through an iterative process. With much less fuss and bother, the empirical weights seem to produce estimates that are as reasonable as the estimates obtained through iteration. I do not know of any theoretical work or simulation to substantiate this, but the work of Berkson (1955) on the minimum logit χ^2 estimators is related supportive evidence. I have had no experience with one step estimators (see Amemiya, 1973), but they may also do as well as the estimates obtained from a full iteration.

Any of the weighted estimators can be disturbed by outliers just as for the unweighted estimators. There has been some theoretical work on robust estimators for heteroscedastic linear models (see Carroll and Ruppert, 1982), but judicious trimming is probably what is mainly used in practice.

An unusual consequence can come from weighting when the intercept α is known to be zero. Consider the heteroscedastic model

$$y_i = \beta x_i + e_i \tag{5.115}$$

with $\text{Var}(e_i) = \sigma_i^2$. The weighted least squares estimate of β is

$$\hat{\beta}_w = \frac{\sum_{i=1}^n w_i x_i y_i}{\sum_{i=1}^n w_i x_i^2} \tag{5.116}$$

where $w_i \propto 1/\sigma_i^2$. For the special case in which $\sigma_i^2 = \sigma^2|\beta|x_i$ with $x_i > 0$, the weight w_i can be taken equal to $1/x_i$. In this event $\hat{\beta}_w$ simplifies to

$$\hat{\beta}_w = \frac{\sum_{i=1}^n y_i}{\sum_{i=1}^n x_i} = \frac{\bar{y}}{\bar{x}}, \tag{5.117}$$

that is, the ratio of the sample means. More generally, as long as σ_i^2 is proportional to a known power of $|\mu(x_i)|$ (i.e., $\sigma_i^2 = \sigma^2|\beta x_i|^\lambda$), the weights are known (i.e., $w_i = |x_i|^{-\lambda}$), and no iteration or empirical weights are required in the estimation procedure.

If the x_i are fixed, or viewed as conditionally fixed, the variability of $\hat{\beta}_w$ is estimated by

$$\widehat{\text{Var}}(\hat{\beta}_w) = \frac{\hat{\sigma}^2}{\sum_{i=1}^n w_i x_i^2} \tag{5.118}$$

where

$$\hat{\sigma}^2 = \frac{1}{n-1}\sum_{i=1}^n w_i(y_i - \hat{\beta}_w x_i)^2. \tag{5.119}$$

As long as the weights w_i are known powers of x_i, the usual normal and χ^2 distribution theory goes through for (5.116) and (5.119), respectively.

Ratios of sample means are treated in Chapter 6. In the context there, the sample mean $\bar{x}$ is viewed as a random variable, and its variability is taken into account in estimating the variation of $\bar{y}/\bar{x}$. Thus, in the special case where x_i is random and $w_i = 1/x_i$, there may be some question over whether to estimate the conditional or unconditional variance of $\hat{\beta}_w$.

5.5. Dependence.

Dependence between the y observations can creep into a regression model in a variety of ways, but the brief discussion here is limited to a few main possibilities.

If the pairs of observations (x_i, y_i) are collected in different blocks, as for example, on different days, from different patients, or with different equipment, one regression line may not adequately model the data. The observations within a block may be more closely related than between blocks and the regression relationship may vary between blocks. Ignoring this *blocking* may still provide nearly unbiased estimates of the regression coefficients if the x values are approximately balanced with regard to blocks, but the estimates of variability can be fouled up.

When blocking is known to be present, the wise statistician investigates whether there is any block effect. This can be accomplished by fitting a separate regression within each block. The actual variability between blocks for the estimated intercepts and slopes can be compared with the average of the variabilities estimated internally from the regressions within the blocks. Formal tests of equality can be executed by computing F statistics or studentized range statistics (see Section 5.1.2).

If the regressions are judged to be different between blocks, it may be more appropriate to think of the slopes and intercepts within blocks as being random effects rather than as fixed effects in Section

5.1.3. This is especially germane when the blocks can be viewed as random effects like days or patients. The statistical model would be

$$y_{ij} = a_i + b_i x_{ij} + e_{ij}, \tag{5.120}$$

$i = 1, \cdots, k$, $j = 1, \cdots, r_i$ $(n = \sum_{i=1}^{k} r_i)$, where the (a_i, b_i) are independent and have a bivariate (normal) distribution, and the e_{ij} are independently [of themselves and the (a_i, b_i)] normally distributed.* The separately estimated $\hat{a}_i$, $\hat{b}_i$, $i = 1, \cdots, k$, can often be treated as independently, identically distributed random variables, and their means and variances estimated by

$$\begin{aligned} \hat{\mu}_a &= \frac{1}{k}\sum_{i=1}^{k} \hat{a}_i, \quad \widehat{\mathrm{Var}}(\hat{\mu}_a) = \frac{1}{k(k-1)}\sum_{i=1}^{k}(\hat{a}_i - \hat{\mu}_a)^2, \\ \hat{\mu}_b &= \frac{1}{k}\sum_{i=1}^{k} \hat{b}_i, \quad \widehat{\mathrm{Var}}(\hat{\mu}_b) = \frac{1}{k(k-1)}\sum_{i=1}^{k}(\hat{b}_i - \hat{\mu}_b)^2. \end{aligned} \tag{5.121}$$

These estimates are especially appropriate when the variation in the σ_i^2 and x_{ij} between blocks can be viewed as random. Weighting the a_i and b_i by their within-block estimates of variability is not recommended. This ignores the variability between blocks and tends to produce unstable estimators (see Section 3.5.4).

When there are two or more regressions to be compared and each contains block effects, the just described procedure of individually estimating the intercepts and slopes within blocks and then treating these estimates as the basic random variables in a multiple sample problem is often very useful.

An extreme form of blocking can occur when *replications* (two or more) are taken at each distinct x value. If the replicate observa-

---

* An appropriate modification should be made to (5.120) and the subsequent analysis if either the slopes or intercepts are judged to be equal between blocks.

tions are taken under the same experimental condition, on the same patient, for example, they may exhibit less variability than the overall variation about the regression line. To treat them as independent observations in a standard analysis may lead to incorrect estimates for the variability of the regression estimators.

To be more specific, consider the case of r replicate observations at each of k distinct x values. The total sample size n equals rk. Let the model be

$$y_{ij} = \alpha + \beta x_i + e_{ij}, \tag{5.122}$$

$i = 1, \cdots, k$, $j = 1, \cdots, r$, where

$$e_{ij} = f_i + g_{ij}. \tag{5.123}$$

The variable f_i is the error for the group of observations at x_i as a whole, and g_{ij} denotes the replication error within the group. The distributional assumptions are that f_i is distributed as $N(0, \sigma_f^2)$, g_{ij} is distributed as $N(0, \sigma_g^2)$, and the f_i and g_{ij} are all independent. The r replicates at x_i are no longer independent because of the common factor f_i representing their communal experimental condition, patient, etc.

If the n observations are (incorrectly) substituted into a standard analysis, the error sum of squares can be written as

$$\begin{aligned}\sum_{i=1}^{k}\sum_{j=1}^{r}(y_{ij} - \hat{\alpha} - \hat{\beta}x_i)^2 &= \sum_{i=1}^{k}\sum_{j=1}^{r}(y_{ij} - \bar{y}_{i\cdot})^2 \\ &\quad + r\sum_{i=1}^{k}(\bar{y}_{i\cdot} - \hat{\alpha} - \hat{\beta}x_i)^2,\end{aligned} \tag{5.124}$$

where the least squares estimates simplify to

$$\begin{aligned}\hat{\alpha} &= \bar{y}_{\cdot\cdot} - \hat{\beta}\bar{x}, \\ \hat{\beta} &= \frac{\sum_{i=1}^{k}(x_i - \bar{x})(\bar{y}_{i\cdot} - \bar{y}_{\cdot\cdot})}{\sum_{i=1}^{k}(x_i - \bar{x})^2}.\end{aligned} \tag{5.125}$$

From (5.124) it should be clear that $\hat{\sigma}^2$ [i.e., $SS(E)/(n-2)$] is estimating

$$\frac{k(r-1)\sigma_g^2 + r(k-2)[\sigma_f^2 + (\sigma_g^2/r)]}{rk-2} \tag{5.126}$$

since $y_{ij} - \bar{y}_{i\cdot} = g_{ij} - \bar{g}_{i\cdot}$ and $\bar{y}_{i\cdot} = \alpha + \beta x_i + f_i + \bar{g}_{i\cdot}$. Expression (5.126) can be rewritten as

$$\sigma^2 = \sigma_g^2 + \sigma_f^2 \frac{r(k-2)}{rk-2}. \tag{5.127}$$

The estimate of the variability in $\hat{\beta}$ [i.e., $\widehat{\mathrm{Var}}(\hat{\beta})$] is trying to estimate

$$\begin{aligned} &\frac{\sigma^2}{\sum_{i=1}^k \sum_{j=1}^r (x_{ij} - \bar{x}_{\cdot\cdot})^2} \\ &\quad = \left[\sigma_g^2 + \sigma_f^2 \frac{r(k-2)}{rk-2}\right] \frac{1}{r\sum_{i=1}^k (x_i - \bar{x})^2}, \\ &\quad = \left[\sigma_f^2 \frac{k-2}{rk-2} + \frac{\sigma_g^2}{r}\right] \frac{1}{\sum_{i=1}^k (x_i - \bar{x})^2}, \end{aligned} \tag{5.128}$$

whereas the actual variance of $\hat{\beta}$ is

$$\left[\sigma_f^2 + \frac{\sigma_g^2}{r}\right] \frac{1}{\sum_{i=1}^k (x_i - \bar{x})^2}. \tag{5.129}$$

Thus the variance σ_f^2 of the block component in $\mathrm{Var}(\hat{\beta})$ is being underestimated by the factor $(k-2)/(rk-2)$.

A similar effect occurs for $\hat{\alpha}$.

The way to detect this is to compare the replication variance estimate

$$\hat{\sigma}^2_{\text{repl}} = \frac{1}{k(r-1)} \sum_{i=1}^k \sum_{j=1}^r (y_{ij} - \bar{y}_{i\cdot})^2 \tag{5.130}$$

with the (suitably scaled) regression variance estimate based on the means $\bar{y}_{i\cdot}$, i.e.,

$$r\hat{\sigma}^2_{\text{reg}} = \frac{r}{k-1}\sum_{i=1}^{k}(\bar{y}_{i\cdot} - \hat{\alpha} - \hat{\beta}x_i)^2. \tag{5.131}$$

If (5.131) is much larger than (5.130), this indicates the presence of a nonzero σ_f^2 in (5.131). Under $H_0 : \sigma_f^2 = 0$ the ratio $r\hat{\sigma}^2_{\text{reg}}/\hat{\sigma}^2_{\text{repl}}$ has an F distribution with $k-1$ and $k(r-1)$ df, but this test is extremely sensitive to the assumption of normality (see Chapter 7).

When there is evidence that the replication variance is smaller than the regression variance, the safest analysis is to average over the replicates and run the regression on the mean values $\bar{y}_{i\cdot}$, $i = 1, \cdots, k$. The incentive for not doing this unless absolutely necessary is the considerable loss in degrees of freedom from $rk - 2$ to $k - 2$.

Another type of dependence can be created by *baseline adjustment*. Although this adjustment procedure may arise in multisample and cross-classification problems as well, it seems to occur more frequently in regression contexts in my experience. It occurs when the recorded observation y_i is actually a measured value u_i that is divided by a baseline measurement u_0 or has a baseline measurement subtracted from it. That is,

$$y_i = u_i/u_0 \quad \text{or} \quad y_i = u_i - u_0, \tag{5.132}$$

$i = 1, \cdots, n$, and these adjusted values y_i are felt to have a linear relationship with the independent variable values x_i, $i = 1, \cdots, n$.

If the investigator has been wise enough to measure u_0 with far greater accuracy than u_i by taking more replicates, observing longer, etc., then for practical purposes u_0 can be considered to be a constant and no dependence is introduced. On the other hand, if $\text{Var}(u_0)$ is nearly the same size as $\text{Var}(u_i)$, then dependence has been created between the y_i by the communal u_0. This needs to be taken into account in the analysis.

With the ratio adjustment $y_i = u_i/u_0$, the estimated regression coefficients of y on x are algebraically equal to the regression coefficients of u on x divided by u_0 (i.e., $\hat{\alpha}_y = \hat{\alpha}_u/u_0$, $\hat{\beta}_y = \hat{\beta}_u/u_0$). Similarly, $\hat{\sigma}^2$ is the mean square error for u regressed on x divided by u_0^2 (i.e., $\hat{\sigma}_y^2 = \sigma_u^2/u_0^2$). The usual tests of $H_0 : \alpha = 0$ and $H_0 : \beta = 0$ are valid, but tests of any nonzero values are conditional on u_0. The customarily estimated variability of $\hat{\alpha}$ and $\hat{\beta}$ does not include a component from the variability in u_0. To incorporate this, the method in Chapter 6 for estimating the variability of a ratio needs to be applied. An estimate of the variance of u_0 is required for this.

With the subtraction adjustment $y_i = u_i - u_0$, the usual estimate of $\hat{\beta}$ and $\hat{\sigma}^2$ are undisturbed since u_0 cancels out, but $\hat{\alpha}_y = \hat{\alpha}_u - u_0$. Clearly, $\text{Var}(\hat{\alpha}_y) = \text{Var}(\hat{\alpha}_u) + \text{Var}(u_0)$. The first component can be estimated from the standard regression analysis, but an estimate of the second has to be obtained elsewhere. If u_0 is in fact an average of a group of m baseline values, $\widehat{\text{Var}}(u_0)$ can be calculated from the sample variance of the group divided by m.

The final type of dependence to be mentioned is *serial correlation*. Concern for this arises most frequently when the index i measures time. The x variable may itself be time or a function of time. The model is still (5.1), but the e_i, which are $N(0, \sigma^2)$, may have $\text{Cov}(e_i, e_{i+j}) = \rho_j \sigma^2 \neq 0$ for $j \geq 1$.

Although the standard regression coefficient estimators remain unbiased, the usual estimates of variability can be quite inaccurate. Durbin and Watson (1950, 1951, 1971) proposed a test for $H_0 : \rho_j \equiv 0$, $j \geq 1$, based on the statistic

$$DW = \frac{\sum_{i=1}^{n-1}(r_{i+1} - r_i)^2}{\sum_{i=1}^{n} r_i^2}, \tag{5.133}$$

where $r_i = \hat{\alpha} - \hat{\beta} x_i$, $i = 1, \cdots, n$ are the residuals from the standard regression analysis. The null hypothesis is rejected for small values of DW. For details see the original sources or Draper and Smith (1981, Chapter 3). Adjustment for serial correlation draws one into the domain of time series analysis, which is beyond the scope of this book.

ERRORS-IN-VARIABLES MODEL

In this model the measurements on both variables are subject to error. Specifically,

$$\begin{aligned} x_i &= u_i + d_i, \\ y_i &= v_i + e_i, \end{aligned} \tag{5.134}$$

where u_i and v_i are the true ith values of the first and second variables, respectively. However, we can only observe x_i and y_i, which are u_i and v_i, respectively, with the observational errors d_i and e_i attached. The true underlying variables u and v are related by the linear relation

$$v = \alpha + \beta u, \tag{5.135}$$

which could just as well be reversed to

$$u = \alpha' + \beta' v, \tag{5.136}$$

where $\beta' = 1/\beta$ and $\alpha' = -\alpha/\beta$. The standard distributional assumptions on the errors are

$$\begin{aligned} d_i &\quad \text{independent} \quad (0, \sigma_d^2), \\ e_i &\quad \text{independent} \quad (0, \sigma_e^2), \\ \{d_i\} &\quad \text{independent of} \quad \{e_i\}. \end{aligned} \tag{5.137}$$

Before launching into a discussion of the statistical analysis associated with the model (5.134)–(5.137), it is important to try to delineate when this model is appropriate and when the already discussed regression model is more valid.

The errors-in-variables model is popular in economic analysis where all the economic variables entering the model are measured with uncertainty. The goal of the analysis for economists is to determine the relationship between the underlying variables.

The errors-in-variables analysis is also especially appropriate when comparing two different techniques for measuring the same quantity, where both techniques experience errors in reproducibility. An example would be measuring cardiac output by the dye-dilution and thermodilution methods; neither method exhibits substantially less variability than the other when repeated measurements are obtained. In this situation, one usually wants to know whether both methods are providing the same reading except for noise, which is the hypothesis $H_0 : \alpha = 0,\ \beta = 1$.

If one of the variables, say x, is the *gold standard* of measurement, then a standard regression analysis may be more appropriate. By "gold standard" is meant that the value of this variable is universally accepted as the true value, even though it may not be any more reproducible or accurate than the other variable. In this circumstance, everyone wants to know how the new variable y relates to the old accepted variable x. When the gold standard measurement is also less variable, there is no question about the analysis; it should be a regression analysis.

When the problem is to predict y from x, the correct analysis is a regression analysis. With mean squared error, $E(y \mid x)$ is the optimal predictor. Under normal theory for u, v, d, and e (see Section

5.6), $E(y \mid x)$ is linear in x, but

$$\begin{aligned} E(y \mid x) &= E(\alpha + \beta u + e \mid x), \\ &= \alpha + \beta E(u \mid x), \\ &= \alpha + \beta \left(\frac{\sigma_d^2}{\sigma_u^2 + \sigma_d^2} \right) \mu_u + \beta \left(\frac{\sigma_u^2}{\sigma_u^2 + \sigma_d^2} \right) x, \end{aligned} \tag{5.138}$$

which differs from $\alpha + \beta x$.* For nonnormal distributions, the last line in (5.138) is still the optimal linear predictor. In either case the least squares regression line is correctly estimating (5.138) [see (5.141)].

Another instance in which regression analysis is the correct analysis is referred to as the *Berkson model* of a *controlled experiment*. Here the first variable is under the control of the investigator, and he or she actually sets the value x_i. However, the true u_i determining v_i may differ from x_i in that the delivered voltage may not equal the set voltage, the drug dilution may not be exact, etc. In these cases where the investigator is setting x_i and not having it measured for him or her, Berkson (1950) showed that one should apply regression analysis.

For a fuller discussion of these issues, the reader is referred to the gold standard reference for this area which is Madansky (1959).

Given that the errors-in-variables model is appropriate, a distinction arises over the character of the u_i (or v_i) values. If they are nonrandom quantities that should be viewed as unknown (design) parameters of the data, then there are $n+4$ parameters in the model to be estimated, namely, u_1 (or v_1), $\cdots$, u_n (or v_n), α (or α'), β (or β'), σ_d^2, and σ_e^2. This submodel is referred to as a *functional relationship* between x and y. The alternative is for the u_i (or v_i) to be viewed as random quantities generated by a distribution with mean μ_u (or μ_v) and variance σ_u^2 (or σ_v^2). The u_i (or v_i) are considered to

* $\mu_u = E(u)$.

be independent of d_i and e_i. This is called a *structural relationship* between x and y. In my opinion, the choice of the words "functional" and "structural" is very unhelpful mnemonically, but we are stuck with them for historical reasons.

The presentation here is limited to the structural relation case. In my experience most applications involve random u_i. There are differences in the analyses, except for the most important special case where $\lambda = \sigma_e^2/\sigma_d^2$ is known. Kendall and Stuart (1961, pp. 383–388) is a good reference for a discussion of the functional relation case.

Before giving the analysis for a structural relation model, it is instructive to see what happens with the usual regression analysis. Consider the slope estimator:

$$\begin{aligned}\hat{\beta} &= \frac{\sum_{i=1}^n (x_i - \bar{x})(y_i - \bar{y})}{\sum_{i=1}^n (x_i - \bar{x})^2}, \\ &= \frac{\sum_{i=1}^n (u_i + d_i - \bar{u} - \bar{d})(\alpha + \beta u_i + e_i - \alpha - \beta \bar{u} - \bar{e})}{\sum_{i=1}^n (u_i + d_i - \bar{u} - \bar{d})^2}, \\ &= \frac{\beta \left[\frac{1}{n}\sum_{i=1}^n (u_i - \bar{u})^2\right] + R_1}{\frac{1}{n}\sum_{i=1}^n (u_i - \bar{u})^2 + \frac{1}{n}\sum_{i=1}^n (d_i - \bar{d})^2 + R_2},\end{aligned} \tag{5.139}$$

where

$$\begin{aligned}R_1 &= \frac{1}{n}\left[\beta \sum_{i=1}^n (d_i - \bar{d})(u_i - \bar{u}) + \sum_{i=1}^n (u_i + d_i - \bar{u} - \bar{d})(e_i - \bar{e})\right], \\ R_2 &= \frac{1}{n}\left[2\sum_{i=1}^n (u_i - \bar{u})(d_i - \bar{d})\right].\end{aligned} \tag{5.140}$$

Since the u_i, d_i, and e_i are independent, R_1 and R_2 converge to zero as $n \to \infty$. Thus

$$\hat{\beta} \xrightarrow{p} \beta\left(\frac{\sigma_u^2}{\sigma_u^2 + \sigma_d^2}\right). \tag{5.141}$$

The usual regression slope estimator is asymptotically biased downward from β by the inaccuracy in the x measurement. The degree of asymptotic inconsistency depends on the relative sizes of σ_d^2 and σ_u^2. However, it is correctly estimating the slope in (5.138). Similarly, $\hat{\alpha}$ is estimating the intercept in (5.138) rather than α.

5.6. Normal Theory.

In addition to the assumption that the $\{d_i\}$, $\{e_i\}$, and $\{u_i$ or $v_i\}$ are all independently distributed, it is postulated that their respective distributions are all normal. This creates a bivariate normal distribution for the pair of observable variables (x_i, y_i).

The *maximum likelihood approach* to estimation of the parameters in the structural errors-in-variables model maximizes the product of the bivariate normal densities for (x_i, y_i), $i = 1, \cdots, n$, with respect to the five parameters μ_x, μ_y, σ_x^2, σ_y^2, and σ_{xy}. The MLE are the usual sample means, variances, and covariance, each with denominator n. Since these five parameters are functions of the six parameters α, β, μ_u, σ_u^2, σ_d^2, and σ_e^2 in the model, the equations relating the bivariate normal parameters to the model parameters define the MLE of the model parameters when estimates are substituted:

$$\begin{aligned} \hat{\mu}_u &= \bar{x}, \\ \hat{\alpha} + \hat{\beta}\hat{\mu}_u &= \bar{y}, \\ \hat{\sigma}_u^2 + \hat{\sigma}_d^2 &= \hat{\sigma}_x^2, \\ \hat{\beta}^2\hat{\sigma}_u^2 + \hat{\sigma}_e^2 &= \hat{\sigma}_y^2, \\ \hat{\beta}\hat{\sigma}_u^2 &= \hat{\sigma}_{xy}, \end{aligned} \tag{5.142}$$

where*,**

* The six parameters α', β', μ_v, σ_v^2, σ_d^2, and σ_e^2 could be used instead; they lead to equations analogous to (5.142).

** Some statisticians may prefer to use the denominator $n - 1$ in $\hat{\sigma}_x^2$, $\hat{\sigma}_y^2$, and $\hat{\sigma}_{xy}$. The estimates $\hat{\alpha}$ and $\hat{\beta}$ are invariant under the choice.

$$\hat{\sigma}_x^2 = \frac{1}{n}\sum_{i=1}^{n}(x_i - \bar{x})^2,$$
$$\hat{\sigma}_y^2 = \frac{1}{n}\sum_{i=1}^{n}(y_i - \bar{y})^2, \qquad (5.143)$$
$$\hat{\sigma}_{xy} = \frac{1}{n}\sum_{i=1}^{n}(x_i - \bar{x})(y_i - \bar{y}).$$

Since (5.142) consists of five equations in six unknowns, there is clearly a difficulty. The parameters are nonidentifiable. Extra information is needed to pick out a unique solution. The sort of information that is usually available involves knowledge about σ_d^2 and σ_e^2. This can take the form that either one or their ratio is known. Typically, replicates have been run for the measurement processes on x and y for previous data or even on the current data. Using this information to establish a value for the relative sizes of the error variances through the ratio $\lambda = \sigma_e^2/\sigma_d^2$ usually produces a more stable result than trying to tightly estimate the absolute size of σ_d^2 or σ_e^2.

When replicates have been run in the experiment, they should be averaged and the average values used in the errors-in-variables analysis. Otherwise, the dependence between different data points is being ignored. Since the data exhibit variability in both the x and y variables, the usual standard error bars, like those in Figure 1.1, should be displayed both horizontally and vertically. This is depicted in Figure 5.9 along with the structural line.

On rare occasions one knows the value for λ without having to separately estimate it. I encountered this once when the measurement processes were identical except for the site of sampling, so λ was necessarily equal to one.

The analysis to follow is restricted to the case of known λ. For

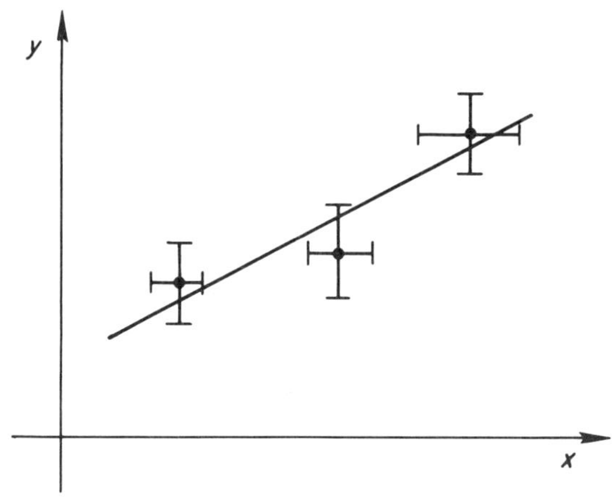

Figure 5.9

a discussion of the other cases, see Kendall and Stuart (1961, pp. 380–382) and Birch (1964).

Since $\hat{\sigma}_d^2$ and $\hat{\sigma}_e^2$ must satisfy $\lambda = \hat{\sigma}_e^2/\hat{\sigma}_d^2$, the term $\lambda\hat{\sigma}_d^2$ can be substituted for $\hat{\sigma}_e^2$ in the fourth equation of (5.142). The third through fifth equations can then be manipulated to eliminate $\hat{\sigma}_u^2$ and $\hat{\sigma}_d^2$. This yields the following quadratic equation for $\hat{\beta}$:

$$\hat{\beta}^2\hat{\sigma}_{xy} + \hat{\beta}\left(\lambda\hat{\sigma}_x^2 - \hat{\sigma}_y^2\right) - \lambda\hat{\sigma}_{xy} = 0. \tag{5.144}$$

The solutions to (5.144) are

$$\hat{\beta} = \frac{\hat{\sigma}_y^2 - \lambda\hat{\sigma}_x^2 \pm \left[\left(\hat{\sigma}_y^2 - \lambda\hat{\sigma}_x^2\right)^2 + 4\lambda\hat{\sigma}_{xy}^2\right]^{1/2}}{2\hat{\sigma}_{xy}}. \tag{5.145}$$

From the last equation in (5.142), $\hat{\beta}$ must have the same sign as $\hat{\sigma}_{xy}$. Since $2\hat{\sigma}_{xy}$ is the denominator in the ratio in (5.145), it follows that the numerator must be positive. This implies that the correct root in (5.145) is the "+" root because the square root term is larger than the preceding term.

The roots of (5.144) are sometimes written in the form

$$\hat{\beta} = U \pm [U^2 + \lambda]^{1/2}, \tag{5.146}$$

where

$$U = \frac{\hat{\sigma}_y^2 - \lambda\hat{\sigma}_x^2}{2\hat{\sigma}_{xy}}. \tag{5.147}$$

The correct root has the same sign as $\hat{\sigma}_{xy}$; that is, if $\hat{\sigma}_{xy} > 0$, use "+," and if $\hat{\sigma}_{xy} < 0$, use "−." See Madansky (1959, Appendix) for errors in the literature on this sign calculation.

Once $\hat{\beta}$ has been determined from (5.145)–(5.147), the estimates of the other parameters follow immediately from the equations (5.142):

$$\begin{aligned} \hat{\mu}_u &= \bar{x}, \\ \hat{\alpha} &= \bar{y} - \hat{\beta}\bar{x}, \\ \hat{\sigma}_u^2 &= \hat{\sigma}_{xy}/\hat{\beta}, \\ \hat{\sigma}_d^2 &= \hat{\sigma}_x^2 - (\hat{\sigma}_{xy}/\hat{\beta}), \\ \hat{\sigma}_e^2 &= \hat{\sigma}_y^2 - \hat{\beta}\hat{\sigma}_{xy}. \end{aligned} \tag{5.148}$$

Although the estimators (5.145) and (5.148) were derived under the assumption of normal distributions, they have a more nonparameteric quality to them. Inspection of the equations (5.142) reveals that they are *method of moments* estimators, where u, d, and e can have any distributions with finite second moments. Also, Deming (1943) and Lindley (1947) have derived these estimators from *weighted least squares* points of view. They are in fact the *orthogonal regression* estimators when the variable scales are appropriately adjusted for σ_d^2 and σ_e^2. Actually, the history of errors-in-variables analysis extends back into the nineteenth century (see Madansy, 1959, Appendix for early references).

When the error variances ratio $\lambda = \sigma_e^2/\sigma_d^2$ is known, the MLE for α, β obtained by maximizing the likelihood with respect to $u_1, \cdots, u_n$ α, β, σ_d^2, and σ_e^2 in the functional errors-in-variables model coincide with those given in (5.145) and (5.148). For details of the derivation, the reader is referred to Kendall and Stuart (1961, pp. 383–386) or Graybill (1961, pp. 189–191).

Additional references for the reader who wishes to delve more deeply into errors-in-variables analysis are Sprent(1966) on a generalized least squares approach, Lindley and El-Sayyad (1968) on a Bayesian approach, T. W. Anderson (1976) on connections with simultaneous equations in econometrics, and Gleser(1981) and Chan and Mak (1983) on a multivariate model.

What about tests and confidence intervals for the unknown parameters in the structural errors-in-variables model? The results here are a little sparse.

Creasy (1956) gave the interval

$$\tan^{-1}\left(\frac{\beta}{\sqrt{\lambda}}\right) \in \tan^{-1}\left(\frac{\hat{\beta}}{\sqrt{\lambda}}\right) \pm \left(\frac{1}{2}\right)\sin^{-1}\left\{2t_{n-2}^{\alpha/2}\left[\frac{\lambda(\hat{\sigma}_x^2\hat{\sigma}_y^2 - \hat{\sigma}_{xy}^2)}{(n-2)((\hat{\sigma}_y^2 - \lambda\hat{\sigma}_x^2)^2 + 4\lambda\hat{\sigma}_{xy}^2)}\right]^{1/2}\right\}, \tag{5.149}$$

where $t_{n-2}^{\alpha/2}$ is the upper $100(\alpha/2)$ percentile of the t distribution with $n-2$ df. This interval has probability nearly $1-\alpha$ of being correct and can be converted into an interval on β. The "nearly" comes from omitting any probability associated with $|\tan^{-1}(\hat{\beta}/\sqrt{\lambda}) - \tan^{-1}(\beta/\sqrt{\lambda})| > \pi/4$ in order to avoid additional roots of a trigonometric equation; for details see the original Creasy reference or Kendall and Stuart (1961, pp. 388–389).

The large sample variances and covariance of $\hat{\alpha}$ and $\hat{\beta}$ can be estimated by

$$
\begin{aligned}
n\widehat{\mathrm{Var}}(\hat{\alpha}) &= \hat{\sigma}_y^2 - 2\hat{\beta}\hat{\sigma}_{xy} + \hat{\beta}^2\hat{\sigma}_x^2 + \frac{\bar{x}^2\hat{\beta}^2}{\hat{\sigma}_{xy}^2}(\hat{\sigma}_x^2\hat{\sigma}_y^2 - \hat{\sigma}_{xy}^2), \\
n\widehat{\mathrm{Var}}(\hat{\beta}) &= \frac{\hat{\beta}^2}{\hat{\sigma}_{xy}^2}(\hat{\sigma}_x^2\hat{\sigma}_y^2 - \hat{\sigma}_{xy}^2), \qquad (5.150) \\
n\widehat{\mathrm{Cov}}(\hat{\alpha}, \hat{\beta}) &= -\frac{\bar{x}\hat{\beta}^2}{\hat{\sigma}_{xy}^2}(\hat{\sigma}_x^2\hat{\sigma}_y^2 - \hat{\sigma}_{xy}^2).
\end{aligned}
$$

The expressions in (5.150) are derived form the influence functions for $\hat{\alpha}$ and $\hat{\beta}$; the details are given in Kelly (1984). Use of (5.150) with normal critical constants will produce approximate confidence intervals and tests. However, these confidence intervals and tests are dependent on the normality assumption; see Section 5.7.

There are no classical methods for obtaining confidence intervals and tests for σ_d^2, σ_e^2, and σ_u^2 (or σ_v^2). This is just as well because whatever they might be would be very sensitive to departures from normality. Jackknifing and bootstrapping the data are the best currently known methods for assessing the variability in the estimates. For related discussion on inference with variances see Chapter 7.

If it is known a priori that $\alpha = 0$, then the nonidentifiability problem vanishes. The $\hat{\alpha}$ term disappears from the first two equations in (5.142) so the estimate of $\hat{\beta}$ is $\bar{y}/\bar{x}$ (i.e., the ratio of the sample means). To assess the variability (and bias) in $\hat{\beta}$ the methods of Chapter 6 can be applied.

When $\alpha = 0$, estimates of σ_d^2, σ_e^2, and σ_u^2 can be obtained from the last three equations in (5.142) as well. For ill-behaved data, it can happen that one of these variance estimates is negative. For example, $\hat{\beta} > 0$ and $\hat{\sigma}_{xy} < 0$ yields $\hat{\sigma}_u^2 < 0$. This means that the actual maximum for the likelihood must occur somewhere on the boundary of the admissible parameter region.

Before closing, a second technique should be briefly mentioned.

It was introduced by Wald (1940) in the context of a functional relationship (i.e., u_i fixed), but there is no reason it cannot be applied in a structural relationship setting. Wald's original idea was to divide the data into two groups according to the ordering on the x-scale. Nair and Shrivastava (1942/44), Nair and Banerjee (1942/44), and Bartlett (1949) generalized this to three groups. Since three groups seem to be more effective than two, the technique is described using three groups.

The *group mean* method orders the data lexicographically according to the values x_i, $i = 1, \cdots, n$, and then divides them into three roughly equally sized groups. An underlying assumption of the method is that the noise terms d_i are small compared to the variation in the u_i so that the data are actually divided into three groups containing the smallest third of the u_i, the middle third, and the largest third. No u_i should be in the wrong group, or, if there are some mistakes, they should be insignificantly few. Let $\bar{x}_1$, $\bar{x}_2$, $\bar{x}_3$ $(\bar{x}_1 < \bar{x}_2 < \bar{x}_3)$ and $\bar{y}_1$, $\bar{y}_2$, $\bar{y}_3$ be the means of the x and y variables in the three groups. Then the group mean estimate of the slope is

$$\hat{\beta} = \frac{\bar{y}_3 - \bar{y}_1}{\bar{x}_3 - \bar{x}_1}, \tag{5.151}$$

and the intercept estimate is

$$\hat{\alpha} = \bar{y} - \hat{\beta}\bar{x}, \tag{5.152}$$

where $\bar{x}$, $\bar{y}$ are the means for all the data. Note that the slope estimate (5.151) ignores the middle third of the data; it also treats the data asymmetrically in that the x variable determines the three groups.

Wald (1940) and Bartlett (1949) considered the distribution theory for $\hat{\alpha}$, $\hat{\beta}$ under the assumption that the d_i and e_i are normally distributed.

5.7. Departures from Assumptions.

Nonlinearity is best spotted though a plot of the data and the estimated linear structural relation. Residuals can be examined as well just as for regression. In this case, one can actually look at residuals in either direction, namely, $y_i - \hat{\alpha} - \hat{\beta}x_i$ or

$$\begin{aligned} x_i - \hat{\alpha}' - \hat{\beta}' y_i &= x_i + \left(\frac{\hat{\alpha}}{\hat{\beta}}\right) - \left(\frac{1}{\hat{\beta}}\right) y_i, \\ &= \left(\frac{-1}{\hat{\beta}}\right)(y_i - \hat{\alpha} - \hat{\beta}x_i). \end{aligned} \tag{5.153}$$

To my knowledge, no formal tests for nonlinearity have been proposed in the errors-in-variables model other than to embed the linear model in a larger model and to test the added parameters for significance. Even this is difficult because of the complexities involved in fitting and testing structural or functional relations for models other than a simple straight line. Some work has been done on the quadratic model $v = \beta_0 + \beta_1 u + \beta_2 u^2$, but it falls outside the intended scope of this book. It is primarily concerned with the functional relation model. For recent work and earlier references, the reader is referred to Wolter and Fuller (1982).

Nonnormality of the underlying d, e, or $u(v)$ distributions is not a serious threat to the accuracy of the estimates of α, β, σ_d^2, σ_e^2, σ_u^2, or σ_v^2 except for the effects of outliers from contaminated or heavy-tailed distributions. The estimates defined by (5.142) are method of moments estimators as well as being maximum likelihood estimators under the normality assumption so their calculation is not dependent on the normality assumption. However, outliers can distort the sample moments entering (5.142) and thus affect the values of the estimators. The position and magnitude of an outlier govern its impact on the estimators just as in regression.

Diagnostics for the errors-in-variables analysis is nowhere as well developed as for regression analysis. Plotting of the data is the best means of spotting outliers. Examination of the residuals $y_i - \hat{\alpha} - \hat{\beta}x_i$ or $x_i - \hat{\alpha}' - \hat{\beta}'y_i$ and their relative positions determined by $(x_i - \bar{x})/[\sum_{i=1}^n (x_i - \bar{x})^2]^{1/2}$ or $(y_i - \bar{y})/[\sum_{i=1}^n (y_i - \bar{y})^2]^{1/2}$ may be useful as well.

At this point in time, judicious trimming is the only antidote used for outliers. Robustics have not yet come to errors-in-variables, but some proposals are given in M. L. Brown (1982).

Although the estimators themselves are modestly robust to a lack of normality (except for outliers), this is not true for their distribution theory. The confidence interval (5.149) and the asymptotic variances (5.150) are very dependent on the normality assumption. This was shown in the work of Kelly (1984), who derived the influence functions for $\hat{\alpha}$ and $\hat{\beta}$ defined by (5.148) and (5.145), respectively, under a general distribution F for (x, y). In a simulation study she compared the estimates of the variability in $\hat{\alpha}$ and $\hat{\beta}$ for the normal theory estimators (5.150) with estimators based on the influence function and the estimators obtained by jackknifing and bootstrapping. The normal theory and influence function estimators performed very poorly. The influence function estimates were consistently much too small. The normal theory estimators also tended to underestimate but by not quite so much. The jackknife and bootstrap did much better with the jackknife always being somewhat conservative. The bootstrap behaved a bit erratically.

Since the *jackknife* yields the preferred estimators of variability at this time, it should be described explicitly. Delete in succession each data point (x_i, y_i), $i = 1, \cdots, n$. With the ith data point deleted, let $\hat{\alpha}_{-i}$ and $\hat{\beta}_{-i}$ be the estimates calculated from (5.145) and (5.148) with the sample moments based on $n - 1$ data points. The pseudo-values are defined by

$$\begin{aligned} \tilde{\alpha}_i &= n\hat{\alpha} - (n-1)\hat{\alpha}_{-i}, \\ \tilde{\beta}_i &= n\hat{\beta} - (n-1)\hat{\beta}_{-i}, \end{aligned} \tag{5.154}$$

for $i = 1, \cdots, n$, where $\hat{\alpha}$ and $\hat{\beta}$ are the estimators based on all n data points. Then the jackknife variance and covariance estimators are

$$\begin{aligned} n\widehat{\mathrm{Var}}_J(\hat{\alpha}) &= \frac{1}{n-1}\sum_{i=1}^{n}(\tilde{\alpha}_i - \tilde{\alpha})^2, \\ n\widehat{\mathrm{Var}}_J(\hat{\beta}) &= \frac{1}{n-1}\sum_{i=1}^{n}(\tilde{\beta}_i - \tilde{\beta})^2, \\ n\widehat{\mathrm{Cov}}_J(\hat{\alpha},\hat{\beta}) &= \frac{1}{n-1}\sum_{i=1}^{n}(\tilde{\alpha}_i - \tilde{\alpha})(\tilde{\beta}_i - \tilde{\beta}), \end{aligned} \tag{5.155}$$

where $\tilde{\alpha} = \sum_{i=1}^{n} \tilde{\alpha}_i/n$ and $\tilde{\beta} = \sum_{i=1}^{n} \tilde{\beta}_i/n$. The expressions in (5.155) are also the variance and covariance estimates for the jackknife estimators $\hat{\alpha}_J = \tilde{\alpha}$ and $\hat{\beta}_J = \tilde{\beta}$. The variability estimates (5.155) used in conjunction with the approximate bivariate normal distribution for $\hat{\alpha} - \alpha$ and $\hat{\beta} - \beta$ (or $\hat{\alpha}_J - \alpha$ and $\hat{\beta}_J - \beta$) yield approximate confidence intervals or regions for α and β.

Brillinger (1966) was the first to suggest applying the jackknife to the linear structural relation problem.

The reader should be warned that the jackknife is not resistant to outliers. Therefore, any influential outliers should be trimmed before jackknifing.

Nothing has been published on what happens or what to do when the observational errors have *unequal variances*. A common situation is where σ_d^2 and σ_e^2 increase as u and v increase. In particular, the observational errors may have a constant coefficient of variation. If a transformation is found that both stabilizes the error variances and maintains or creates a linear relation between u and v, the problem is resolved, but most of the time we are not so lucky.

Dependence between data points in the errors-in-variables model – who knows anything?

Exercises.

1. Verify that the expressions in (5.10) are the variances-covariance for $\hat{\alpha}$, $\hat{\beta}$ for fixed $x_1, \cdots, x_n$ and independently, identically (but not necessarily normally) distributed $e_1, \cdots, e_n$.
2. Verify the relationship between r and $\hat{\beta}/\widehat{SD}(\hat{\beta})$ given in (5.22).
3. Use the probability equality

$$P\{\alpha^* \in \hat{\alpha}^* \pm |m|^{\alpha}_{2,n-2}\hat{\sigma}(1/n)^{1/2}$$

and

$$\beta^* \in \hat{\beta}^* \pm |m|^{\alpha}_{2,n-2}\hat{\sigma}(1/S_{xx})^{1/2}\} = 1 - \alpha,$$

to prove that

$$P\left\{\alpha + \beta x \in \hat{\alpha} + \hat{\beta}x \pm |m|^{\alpha}_{2,n-2}\hat{\sigma}\left(\frac{1}{n^{1/2}} + \frac{|x - \bar{x}|}{S_{xx}^{1/2}}\right) \text{ for all } x\right\} = 1 - \alpha,$$

where $S_{xx} = \sum_{i=1}^{n}(x_i - \bar{x})^2$.

Hint: Figure out the projections of the confidence rectangle onto $(1, x - \bar{x})$, or, for $P\{A\}$ and $P\{B\}$ above, show that $A \subset B$ and $A \supset B$.

4. Use the delta method to show that asymptotically

$$\mathrm{Var}\left(\frac{\mu_0 - \hat{\alpha}}{\hat{\beta}}\right) = \frac{\sigma^2}{\beta^2}\left[\frac{1}{n} + \frac{(\mu_0 - \alpha - \beta\bar{x})^2}{\beta^2 S_{xx}}\right]$$

as $n \to \infty$, $S_{xx} \to \infty$ [see expression (5.35)].

Hint: Use the Taylor series expansion

$$g(u,v) = g(\mu,\eta) + (u-\mu)\frac{\partial}{\partial u}g(\mu,\eta) + (v-\eta)\frac{\partial}{\partial v}g(\mu,\eta) + \cdots,$$

where $\mu = E(u)$ and $\eta = E(v)$.

5. For the multisample regression problem with common slope, obtain an expression for Var $\left(\sum_{i=1}^{I} c_i\hat{\alpha}_i\right)$, where the intercept estimates $\hat{\alpha}_i$, $i = 1,\cdots,I$, are defined by (5.63) and c_i, $i = 1,\cdots,I$, constitute a contrast.

6. Use the delta method to justify the bias-adjusted estimator $\hat{\hat{\Delta}}_{12}$ given in (5.75).

7. Verify the value of $E(y \mid x)$ given in (5.138) for the normal theory errors-in-variables structural model.

8. For the errors-in-variables structural model with known error variances ratio $\lambda = \sigma_e^2/\sigma_d^2$, derive equation (5.144) and the estimates (5.148) from the relations in (5.142).

9. In the data in Exercise 6 of Chapter 1, donor blood was collected into paired bags containing ACD and ACD+A.

   (a) Compute the product moment correlation coefficient r for the 12 paired values.

   (b) Does the size of r have any implication for whether a paired or unpaired t test should be run?

   (c) Is r statisticlaly significantly different from zero?

   (d) Compute Kendall's rank correlation coefficient $\hat{\tau}$.

   (e) Is $\hat{\tau}$ statistically significantly different from zero?

10. For the data in Exercise 9 of Chapter 3, regress the girls' muscle grades on their (group) ages. Is the variability in the 5 mean values about the regression line consistent with the variabilty within age groups?

11. The 6 commercially available contrast agents used in the experiment described in Exercise 10 of Chapter 4 are chemically different. However, the investigator felt that the only quantity that affected the opacification index was the amount of iodine in each agent. The iodine concentrations (mg I/ml) in the agents are as follows:

$$A = 400 \qquad B = 320 \qquad C = 400$$

$$D = 480 \qquad E = 370 \qquad F = 282$$

A full analysis of the extra period Latin square data indicates that there are no period or residual effects, but there are dog effects. For the purpose of this exercise discard the data from the extra period.

   (a) Calculate a linear regression between opacification index and iodine concentration. Obtain estimated standard deviations for the regression coefficient estimates.

   (b) Is the investigator justified in his claim that the iodine concentration in the agent determines the opacification index?

   (c) Is there any evidence of nonlinearity between the opacification indices and iodine concentrations?

12. Blood volume in a newborn can be calculated by injecting dye and then dividing the amount injected by the concentration mesured in the blood. The optical density (OD) of the dye is measured at wavelength 620. However, other color agents in the blood have density curves overlapping this wavelength. Because of this it is necessary for the blood volume calculation to subtract from the optical density at wavelength 620 an estimate of the density without dye predicted from the density measured at wavelength 740, which is unaffected by the dye.

The optical densities (without dye) at wavelengths 620 and 740 for 36 newborns on phototherapy for jaundice are presented in the table. A prediction equation from which the OD 620 can be estimated from the OD 740 is desired.

(a) Obtain the regression estimates $\hat{\alpha}$, $\hat{\beta}$, and $\hat{\sigma}^2$.

(b) Is OD 740 significantly related to OD 620?

(c) Does any adjustment need to be made for departures from assumptions?

(d) What is your final prediction equation?

Newborn	Od 620	OD 740	Newborn	OD 620	OD 740
01	28	14	19	26	9
02	14	7	20	36	17
03	37	12	21	48	20
04	84	40	22	54	30
05	28	11	23	56	31
06	38	16	24	135	74
07	98	54	25	40	16
08	21	9	26	21	8
09	44	22	27	48	19
10	118	74	28	30	10
11	42	18	29	22	11
12	60	31	30	50	30
13	106	48	31	18	8
14	62	42	32	35	16
15	49	22	33	241	124
16	38	18	34	73	29
17	26	9	35	40	11
18	46	23	36	42	20

13. Premature babies are extremely susceptible to infections. At the Stanford Medical Center Intensive Care Nursery, kanamycin is used for the treatment of sepsis. Because kanamycin is ineffective at low levels and has potentially harmful side effects at high levels, it is necessary to constantly monitor its level in the blood. The standard procedure is to take blood samples from a baby's heel. Unfortunately, frequent samples leave badly bruised heels.

    Kanamycin is routinely administered through an umbilical catheter. An alternative procedure to a heelstick for measuring the serum kanamycin level is to reverse the flow in the catheter and draw a blood sample from it. However, physicians are reluctant to rely on measurements from the catheter because proximity to the site of infusion and residual amounts of kanamycin on the wall of the catheter might elevate the levels.

A study of 20 premature babies was conducted to see if kanamycin levels measured in blood drawn from the heel and the catheter are equivalent.* The data from simultaneously drawn samples are presented in the table. Since the preparation and assay processs are identical for both blood samples, it is reasonable to suppose that $\lambda = \sigma_e^2/\sigma_d^2$ equals 1 in a structural errors-in-variables model.

(a) Estimate the intercept α and slope β in a structural errors-in-variables model by normal theory maximum likelihood under the assumption $\lambda = 1$.

(b) Estimate the variability in $\hat{\alpha}$ and $\hat{\beta}$ by the jackknife method.

(c) Would you conclude that the two methods are equivalent?

---

* For additional details on this study see Miller, R. G., Jr. (1980), *Kanamycin levels in premature babies*, Biotatistics Casebook, Vol. 3 (Technical Report No. 57, Division of Biostatistics, Stanford University), pp. 127–142.

Baby	Heelstick	Catheter
01	23.0	25.2
02	33.2	26.0
03	16.6	16.3
04	26.3	27.2
05	20.0	23.2
06	20.0	18.1
07	20.6	22.2
08	18.9	17.2
09	17.8	18.8
10	20.0	16.4
11	26.4	24.8
12	21.8	26.8
13	14.9	15.4
14	17.4	14.9
15	20.0	18.1
16	13.2	16.3
17	28.4	31.3
18	25.9	31.2
19	18.9	18.0
20	13.8	15.6

Chapter 6

RATIOS

For two variables x and y, interest may center on the ratio of their means η/μ, where $\mu = E(x)$ and $\eta = E(y)$. The mean ratio is pertinent when one wants to know the average amount of variable y per unit of variable x. An example might be the protein content of cells per unit DNA. The mean ratio is also involved when the parameter of interest is the percentage change or relative change between experimental (y) and control (x) conditions. *Percentage change* is defined by

$$\left(\frac{\eta - \mu}{\mu}\right) \times 100 = \left(\frac{\eta}{\mu} - 1\right) \times 100, \tag{6.1}$$

and *relative change* is defined by (6.1) without the factor 100.

For a sample of pairs (x_i, y_i), $i = 1, \cdots, n$, an obvious estimator of η/μ is $\bar{y}/\bar{x}$, where $\bar{x}$ and $\bar{y}$ are the sample means. This estimator, which is the ratio of the sample means, will converge to η/μ as $n \to \infty$, but the estimator $(1/n)\sum_{i=1}^{n}(y_i/x_i)$, which is the mean of the sample ratios, will not. The latter estimator is consistent for the expectation $E(y/x)$. The two quantities $E(y/x)$ and $E(y)/E(x)$ are rarely equal. They can be nearly equal or quite far apart, as determined by the joint distribution of x and y. The investigator and statistician should have clearly in mind which quantity they want to estimate. In most cases it is the ratio of the population means η/μ.

The ratio of the means estimator $\bar{y}/\bar{x}$ came to the fore twice in Chapter 5, once in the regression model as a weighted least squares estimator when $\alpha = 0$, and once in the errors-in-variables model as the MLE of β when $\alpha = 0$. The values x_i were asumed to be fixed in the regression model, but the probability structure for $\bar{y}/\bar{x}$ in the structural errors-in-variables model was basically the same as that considered in this chapter.

Ratio estimators play an important role in survey sampling. However, the context in which they are used in sample surveys is a bit different from that being considered here. The probability framework frequently postulated for survey work is sampling without replacement from a finite population of N units. Moreover, often the objective is to estimate $\bar{Y}$, the finite population mean of the y variable. The finite population mean $\bar{X}$ may be known from a more complete survey or census (possibly obtained previously), or it may be known through routine tabulation of other population statistics. When $\bar{X}$ is known, the ratio-type estimate $(\bar{y}/\bar{x})\bar{X}$ can be a more acurate estimate of $\bar{Y}$ than the simple estimate $\bar{y}$.

In survey sampling the ratio estimator has been generalized to stratified samples and other more complex sampling schemes. An excellent reference on the use of ratios in sample surveys is Cochran (1977, Chapter 6). With minor modification from finite to infinite population sampling, most of the discussion in Cochran's book applies to the situation being considered in this chapter.

6.1. Normal Theory.

The one sample problem with *paired data* is the primary focus of attention. Let (x_i, y_i), $i = 1, \cdots, n$, be n independent pairs of values that are distributed according to a bivariate normal distribution with mean vector and covariance matrix given by

$$\begin{pmatrix} \mu \\ \eta \end{pmatrix} \quad \text{and} \quad \begin{pmatrix} \sigma_x^2 & \sigma_{xy} \\ \sigma_{xy} & \sigma_y^2 \end{pmatrix}, \tag{6.2}$$

respectively. The problem is to test whether $\theta = \eta/\mu$ has a specified value θ_0, or to construct a confidence interval for θ.

The *maximum likelihood estimator* of θ is $\hat{\theta} = \bar{y}/\bar{x}$. This is an obvious and intuitive estimator, but it has some difficulties associated with it. For one, it is not an unbiased estimator. In fact, its expectation does not even exist in an absolute sense. For sampling from distributions other than the normal, its expectation can exist, and in the next section modified estimators that reduce the bias in small samples are discussed. For another, the distribution of $\bar{y}/\bar{x}$ is exceedingly complicated and unsuited to confidence interval construction or hypothesis testing (see Geary, 1930; Fieller, 1932; Marsaglia, 1965; and Hinkley, 1969). However, a trick allows one to construct confidence intervals and test hypotheses.

Paulson (1942) explicitly described the following procedure for constructing a confidence interval on $\theta = \eta/\mu$. Earlier Fieller (1940) had used the same idea in a regression context [see Section 5.1.1, (5.31)-(5.33)]. Precursors of the procedure appear in Geary (1930) and Fieller (1932).

Under the bivariate normal model, the variables $z_i = y_i - \theta x_i$, $i = 1, \cdots, n$, are independently normally distributed with mean

$$\eta - \theta\mu = \eta - \left(\frac{\eta}{\mu}\right)\mu = 0 \tag{6.3}$$

and variance

$$\sigma_y^2 - 2\theta\sigma_{xy} + \theta^2\sigma_x^2. \tag{6.4}$$

Consequently, the ratio

$$\frac{\bar{z}}{\left[\frac{1}{n(n-1)}\sum_{i=1}^{n}(z_i - \bar{z})^2\right]^{1/2}} \tag{6.5}$$

has a t distribution with $n - 1$ df. In terms of the original variables,

(6.5) can be written as

$$\frac{\bar{y}-\theta\bar{x}}{\left[\frac{1}{n}\left(s_y^2-2\theta s_{xy}+\theta^2 s_x^2\right)\right]^{1/2}}, \tag{6.6}$$

where s_x^2, s_y^2, and s_{xy} are the sample variances and covariance for the x, y variables.*

The $100(1-\alpha)\%$ confidence region for η/μ consists of all values θ for which the absolute values of (6.6) does not exceed $t_{n-1}^{\alpha/2}$, or equivalently, for which the square of (6.6) does not exceed $F_{1,n-1}^{\alpha}$. In most instances the confidence region has upper and lower limits (θ_U and θ_L, respectively) which are the roots of the quadratic equation

$$(\bar{y}-\theta\bar{x})^2=\frac{1}{n}F_{1,n-1}^{\alpha}\left(s_y^2-2\theta s_{xy}+\theta^2 s_x^2\right). \tag{6.7}$$

This can be rearranged to

$$\theta^2\left(\bar{x}^2-F_{1,n-1}^{\alpha}\frac{s_x^2}{n}\right)-2\theta\left(\bar{x}\bar{y}-F_{1,n-1}^{\alpha}\frac{s_{xy}}{n}\right) + \left(\bar{y}^2-F_{1,n-1}^{\alpha}\frac{s_y^2}{n}\right)=0. \tag{6.8}$$

Thus

$$\theta_L,\theta_U=\frac{\bar{x}\bar{y}-F_{1,n-1}^{\alpha}\frac{s_{xy}}{n}}{\bar{x}^2\left(1-F_{1,n-1}^{\alpha}\frac{s_x^2}{n\bar{x}^2}\right)} \pm\frac{\left(F_{1,n-1}^{\alpha}\right)^{1/2}\left[\frac{\bar{x}^2}{n}\left(s_y^2-2\hat{\theta}s_{xy}+\hat{\theta}^2 s_x^2\right)-F_{1,n-1}^{\alpha}\left(\frac{s_x^2 s_y^2-s_{xy}^2}{n^2}\right)\right]^{1/2}}{\bar{x}^2\left(1-F_{1,n-1}^{\alpha}\frac{s_x^2}{n\bar{x}^2}\right)}. \tag{6.9}$$

If the quantity inside the radical in (6.9) should turn out to be negative, then the roots of the quadratic equation are imaginary, and

* $s_x^2=\sum_{i=1}^{n}(x_i-\bar{x})^2/(n-1)$, $s_y^2=\sum_{i=1}^{n}(y_i-\bar{y})^2/(n-1)$, and $s_{xy}=\sum_{i=1}^{n}(x_i-\bar{x})(y_i-\bar{y})/(n-1)$.

the entire real line constitutes the confidence interval. Also, if $\hat{\theta}$ does not lie between θ_U and θ_L, then the confidence region consists of the whole real line except for the points between θ_U and θ_L, i.e., two semi-infinite intervals. These latter two pathologies are not apt to occur unless the sample size is unusually small and the data are particularly variable.*

When n is large, the higher order terms in n become negligible relatively, so the interval simplifies to

$$\hat{\theta} \pm t_{n-1}^{\alpha/2} \frac{1}{|\bar{x}|}(s_y^2 - 2\hat{\theta}s_{xy} + \hat{\theta}^2 s_x^2)^{1/2}. \tag{6.10}$$

Since $(F_{1,n-1}^{\alpha})^{1/2} = t_{n-1}^{\alpha/2}$, the more natural t critical constant is used in (6.10). The quantity multiplying the t critical constant in (6.10) is the large sample standard devation of $\hat{\theta} = \bar{y}/\bar{x}$, which one could also obtain through the delta method.

To test the hypothesis $H_0 : \theta = \theta_0$ that θ has a specified value θ_0, one can check whether θ_0 lies in the confidence region, or, if the roots (6.9) have not been computed, one can simply compute (6.6) with θ_0 substituted for θ and check whether the absolute value of the ratio exceeds $t_{n-1}^{\alpha/2}$. If the specified value is $\theta_0 = 1$, this procedure reduces to the ordinary one sample t test of the mean equaling zero for paired differences.

Multiple ratios can arise in two contexts.

The first is where more than two variables are measured. Here there are p-dimensional vectors of observations $\boldsymbol{y}$ and means $\boldsymbol{\mu}$, and the objects of interest are the ratios of the mean corrdinates μ_i/μ_j,

---

* These two cases arise when the quadratic inequality $a\theta^2 + b\theta + c \leq 0$ defining the confidence region (i.e., (6.8) with ≤ 0) has $b^2 - 4ac < 0$ (imaginary roots) and $b^2 - 4ac > 0$, $a < 0$ (two semi-infinite lines). The point $\bar{y}/\bar{x}$ is always in the region because at it the left-hand side of (6.7) is zero and the inequality is satisfied.

$i, j = 1, \cdots, p$. This leads into the arena of multivariate analysis. Scheffé (1970b) investigated the construction of simultaneous confidence intervals for all ratios of interest

The second context in which multiple ratios arise is in the comparison of ratios for different populations. For population i, $i = 1, \cdots, I$, the ratio $\hat{\theta}_i = \bar{y}_i / \bar{x}_i$ estimates the mean ratio $\theta_i = \eta_i / \mu_i$. Interest may center on the equality of $\theta_1, \cdots, \theta_I$, or the lack thereof. No special procedures have been developed for handling this problem. If the samples are not small, then the $\hat{\theta}_i$ are approximately independently normally distributed with means θ_i and variances

$$\frac{1}{\mu_i^2}(\sigma_{yi}^2 - 2\theta_i \sigma_{xyi} + \theta_i^2 \sigma_{xi}^2). \tag{6.11}$$

Sample moment estimates can be substituted for the unknown parameters in (6.11). Ad hoc test procedures and confidence intervals can sometimes be created on this basis, but inequality of the variances (6.11) for different i is a thorn in one's side. The large sample model structure is essentially that of a one-way classification with unequal variances between populations so the discussion in Section 3.3 is relevant.

Malley (1982) considered the multiple comparisons aspects of both types of multiple ratio problems and their combination.

Although paired data are more common for ratio problems, the case of *unpaired data* does arise on occasion. In this case the x_i, $i = 1, \cdots, m$, are assumed to be independently distributed as $N(\mu, \sigma_x^2)$ and the y_i, $i = 1, \cdots, n$, as independent $N(\eta, \sigma_y^2)$ variables. The x_i and y_i are also assumed to be independent. The two sample sizes m and n need not be equal as with paired data.

The *maximum likelihood estimator* of $\theta = \eta/\mu$ is $\hat{\theta} = \bar{y}/\bar{x}$ just as with paired data. If it is possible to assume that the two population variances are equal (i.e., $\sigma_x^2 = \sigma_y^2 = \sigma^2$), then the confidence interval and test procedure closely parallel those for paired data with zero correlation and are based on exact distribution theory.

Let s^2 be the pooled estimate of variance with $m+n-2$ df, i.e.,

$$\begin{aligned} s^2 &= \frac{(m-1)s_x^2 + (n-1)s_y^2}{m+n-2}, \\ &= \frac{1}{m+n-2}\left[\sum_{i=1}^{m}(x_i-\bar{x})^2 + \sum_{i=1}^{n}(y_i-\bar{y})^2\right]. \end{aligned} \tag{6.12}$$

The $100(1-\alpha)\%$ confidence interval consists of all values θ for which the ratio

$$\frac{\bar{y}-\theta\bar{x}}{s\sqrt{\frac{1}{n}+\frac{\theta^2}{m}}} \tag{6.13}$$

does not exceed the critical constant $t_{m+n-2}^{\alpha/2} = (F_{1,m+n-2}^{\alpha})^{1/2}$ in absolute value. The values of θ where the ratio actually equals the critical constant are the roots of the quadratic equation

$$\theta^2\left(\bar{x}^2 - F_{1,m+n-2}^{\alpha}\frac{s^2}{m}\right) - 2\theta(\bar{x}\bar{y}) + \left(\bar{y}^2 - F_{1,m+n-2}^{\alpha}\frac{s^2}{n}\right) = 0, \tag{6.14}$$

which are

$$\frac{\bar{x}\bar{y} \pm (F_{1,m+n-2}^{\alpha})^{1/2} s\left[\bar{x}^2\left(\frac{1}{n}+\frac{\hat{\theta}^2}{m}\right) - F_{1,m+n-2}^{\alpha}\frac{s^2}{mn}\right]^{1/2}}{\bar{x}^2\left(1 - F_{1,m+n-2}^{\alpha}\frac{s^2}{m\bar{x}^2}\right)}. \tag{6.15}$$

The confidence interval for θ is the interval between the two roots (6.15) except when the following two pathologies occur. If the quantity inside the radical in (6.15) is negative, then the entire real line constitutes the confidence interval. If $\hat{\theta} = \bar{y}/\bar{x}$ does not lie between the two roots, then the confidence region consists of all values below the lower root and above the upper root. Neither of these oddities is apt to happen unless s is large relative to m, n, and $\bar{x}$.

For large sample sizes the roots (6.15) are approximately

$$\frac{\bar{y}}{\bar{x}} \pm t^{\alpha/2}_{m+n-2} \frac{s}{|\bar{x}|} \left(\frac{1}{n} + \frac{\hat{\theta}^2}{m} \right)^{1/2} . \tag{6.16}$$

The quantity multiplying the t critical constant in (6.16) is the delta method estimate of the standard deviation of $\bar{y}/\bar{x}$. The square of it [i.e., $\widehat{SD}^2(\bar{y}/\bar{x})$] can be written in the form

$$\frac{\widehat{SD}^2(\bar{y}/\bar{x})}{(\bar{y}/\bar{x})^2} = \frac{\widehat{SD}^2(\bar{x})}{\bar{x}^2} + \frac{\widehat{SD}^2(\bar{y})}{\bar{y}^2}, \tag{6.17}$$

where $\widehat{SD}(\bar{x}) = s/\sqrt{m}$ and $\widehat{SD}(\bar{y}) = s/\sqrt{n}$. Expression (6.17) is easily remembered as

$$\widehat{CV}^2(\bar{y}/\bar{x}) = \widehat{CV}^2(\bar{x}) + \widehat{CV}^2(\bar{y}), \tag{6.18}$$

where $CV(z)$ denotes the coefficient of variation of z [i.e., $CV(z) = SD(z)/E(z)$]. For a ratio of independent means the squares of the coefficients of variation add (approximately) as in (6.18). This holds true even if the variances of x and y are different.*

To test the hypothesis $H_0 : \theta = \theta_0$ the ratio

$$\frac{\bar{y} - \theta_0 \bar{x}}{s\sqrt{\frac{1}{n} + \frac{\theta_0^2}{m}}} \tag{6.19}$$

can be compared with a t_{m+n-2} percentile or can be used to calculate a P value from the same table. When the hypothesized value θ_0 is 1, the statistic (6.19) reduces to the usual two sample t statistic for unpaired data [see (2.1)].

When it is not possible to assume that the variances for x and y are equal (i.e., $\sigma_x^2 \neq \sigma_y^2$), difficulties similar to those encountered for the two sample t statistic arise. See the next section for discussion.

* A similar interpretation can be given to (6.11) if a coefficient of covariation is defined.

Multiple ratios involving independent samples may present themselves for analysis. They must be dealt with on an ad hoc basis. No schemata for their analysis have been written down anywhere.

6.2. Departures from Assumptions.

The bivariate normal is not the most satisfactory distribution for modeling paired data in problems involving the ratios of means. Most data in mean ratio problems are positive and possibly skewed toward higher values, wheres the bivariate normal is symmetric along its principal axes and has infinitely long tails in all directions. In many other types of problems the infinitesimal probability on the negative part of the axis does not cause any difficulty but here it does. Expectations, variances, mean squared errors, etc., for ratios fail to exist in the absolute (Lebesgue) sense because of the density where the denominator can be close to zero. There are methods of circumventing this, but the difficulty is a nuisance.

Alternative models assume that x and y are related by

$$y_i = \alpha + \beta x_i + e_i, \tag{6.20}$$

where $E(e_i \mid x_i) = 0$ and the conditional variance is allowed to depend on x through a power relationship, i.e.,

$$\operatorname{Var}(e_i \mid x_i) = \delta |x_i|^{\lambda}, \tag{6.21}$$

where $\delta,\ \lambda > 0$. This permits the fan-shaped behavior displayed in Figure 5.8. The marginal distribution of x is usually assumed to be either a gamma distribution or a log normal distribution.* Both these distributions are skewed to the right.

Much of the research on ratio estimation has been concentrated on reducing the bias in the estimator $\bar{y}/\bar{x}$ created by the fact that

---

* The random variable x has a log normal distriution if $\log x$ is normally distributed.

for most distributions $E(x^{-1}) \neq \mu^{-1}$. A variety of estimators have been proposed of which three are applicable to the situation being considered in this chapter.

Beale's (1962) *estimator* is

$$\hat{\theta}_B = \frac{\bar{y}}{\bar{x}}\left[\frac{1 + (s_{xy}/n\bar{x}\bar{y})}{1 + (s_x^2/n\bar{x}^2)}\right]. \tag{6.22}$$

Tin's (1965) *estimator* is

$$\hat{\theta}_T = \frac{\bar{y}}{\bar{x}}\left[1 + \frac{1}{n}\left(\frac{s_{xy}}{\bar{x}\bar{y}} - \frac{s_x^2}{\bar{x}^2}\right)\right]. \tag{6.23}$$

When $E(\bar{y}/\bar{x})$ exists absolutely under the model, the delta method gives

$$\begin{aligned}\frac{\bar{y}}{\bar{x}} &= \frac{\eta}{\mu} + (\bar{x} - \mu)\left(\frac{-\eta}{\mu^2}\right) + (\bar{y} - \eta)\left(\frac{1}{\mu}\right) \\ &+ \frac{1}{2}\left[(\bar{x} - \mu)^2\left(\frac{2\eta}{\mu^3}\right) + 2(\bar{x} - \mu)(\bar{y} - \eta)\left(\frac{-1}{\mu^2}\right) + (\bar{y} - \eta)^2(0)\right] \\ &+ o(\max\{(\bar{x} - \mu)^2, (\bar{y} - \eta)^2\}),\end{aligned} \tag{6.24}$$

so

$$\begin{aligned}E\left(\frac{\bar{y}}{\bar{x}}\right) &= \frac{\eta}{\mu} + \frac{\sigma_x^2}{n}\left(\frac{\eta}{\mu^3}\right) - \frac{\sigma_{xy}}{n}\left(\frac{1}{\mu^2}\right) + o\left(\frac{1}{n}\right), \\ &= \frac{\eta}{\mu}\left[1 + \frac{\sigma_x^2}{n\mu^2} - \frac{\sigma_{xy}}{n\mu\eta} + o\left(\frac{1}{n}\right)\right].\end{aligned} \tag{6.25}$$

The Beale and Tin estimators, each in its own way, are clearly designed to eliminate the $1/n$ bias term in (6.25).

The third estimator, the *jackknife*, was introduced by Quenouille (1949), to reduce the bias of a serial correlation estimator. Quenouille (1956) briefly considered the reciprocal of a mean, but it was Durbin (1959) who first studied in detail the application of the jackknife to ratio problems. If the data are randomly divided into two groups of size $n/2$ (assumed to be an integer), the jackknife is defined by

$$\hat{\theta}_J = 2\left(\frac{\bar{y}}{\bar{x}}\right) - \frac{1}{2}\left(\frac{\bar{y}_1}{\bar{x}_1} + \frac{\bar{y}_2}{\bar{x}_2}\right), \tag{6.26}$$

where $\bar{x}_i, \bar{y}_i$ are the means for the ith group. However, intuition and papers by J. Rao (1965) and J. Rao and Webster (1966) indicate that the full jackknife (see Section 5.7) defined by

$$\hat{\theta}_J = n\hat{\theta} - (n-1)\left(\frac{1}{n}\sum_{i=1}^{n}\hat{\theta}_{-i}\right), \tag{6.27}$$

where $\hat{\theta}_{-i} = \bar{y}_{-i}/\bar{x}_{-i}$ is the ratio with the ith pair (x_i, y_i) deleted from the data, constitutes an improvement, albeit slight.

When the sampling is from a finite population and the mean $\bar{X}$ of the x variable is known, there exist unbiased estimators proposed by Hartley and Ross (1954) and Mickey (1959), and an approximately unbiased estimator by Nieto de Pascual (1961), but these are not applicable to the problem being considered here.

There have been a number of papers comparing various subsets of these estimators under sampling from the bivariate normal model (6.2), the regression model (6.20) and (6.21) with x either gamma or log normal, and a selection of actual finite populations. The list includes Tin(1965), J. Rao and Beegle (1967), J. Rao (1969), P. Rao (1969), Hutchinson (1971), P. Rao and J. Rao (1971) as well as others.

Although the estimators $\hat{\theta}_B$, $\hat{\theta}_T$, and $\hat{\theta}_J$ have smaller bias than $\hat{\theta}$, they tend to have increased variability. Since the mean squared error is the sum of the variance and the square of the bias, the effect on overall performance is unclear. The Monte Carlo and theoretical results that have been obtained about the MSE are confusing. The superiority or inferiority of any one estimator seems to depend on the particular sampling model. In general, $\hat{\theta}_B$, $\hat{\theta}_T$, and $\hat{\theta}_J$ tend to do a bit better than $\hat{\theta}$, but any difference is very slight if it exists at all. The variabilities in the three bias-reduced estimators increase relative to the unadjusted ratio estimator as λ in (6.21) increases. There is little to choose between $\hat{\theta}_B$, $\hat{\theta}_T$, and $\hat{\theta}_J$ as they all perform essentially the same. The jackknife estimator requires more computation, which is a disadvantage, and it is more erratic for small samples.

My impression from having tried the bias-reduced estimators in a few biostatistical problems is that the correction for bias is usually relatively very small in magnitude. The amount of change is of no consequence to the investigator. This would seem to be in agreement with the previously cited studies. However, it should be remarked that the studies indicate improvement should occur with the use of bias-reduced estimators when there are multiple strata in finite population sampling.

Less attention has been paid to the skewness and kurtosis of $\hat{\theta} = \bar{y}/\bar{x}$ and the three bias-reduced estimators. A few results are mentioned in J. Rao (1969). The asymptotic normality of $\hat{\theta}$ for any distribution of (x, y) with finite second moments is guaranteed by the bivariate central limit theorem for $(\bar{x}, \bar{y})$ and the asymptotic normality of a continuously differentiable function of sample means (see C. R. Rao, 1973, Section 6a.2). Scott and Wu (1981) establish the asymptotic normality under finite population sampling. For small to moderately sized samples there is some evidence that for positive x, y the distribution of $\bar{y}/\bar{x}$ is positively skewed. This skewness is primarily caused by small values of $\bar{x}$.

The statistic (6.6) is the basis for testing hypotheses and constructing confidence intervals. It is a one sample t statistic (6.5). Thus the effects of skewness, kurtosis, and outliers on tests and confidence intervals are similar to those mentioned in Section 1.2.1.

There has been no work to date on robust estimation procedures especially designed for ratio estimation to counter the effects of heavy-tailed distributions.

Several papers (viz., J. Rao, 1969, and P. Rao and J. Rao, 1971) have indicated that the estimate

$$\frac{1}{\bar{x}^2}(s_y^2 - 2\hat{\theta}s_{xy} + \hat{\theta}^2 s_x^2) \tag{6.28}$$

of the variance used in (6.10) is a biased estimate. It can be biased positively or negatively depending on the model. However, it is a consistent estimate of the correct asymptotic variance for any underlying distribution with finite second moments. Therefore, in large samples the interval (6.10) must be correct even if the distribution for x and y is not a bivariate normal.

An alternative estimate of the variance is provided by the jackknife estimate of variance (see Section 5.7). Its application to ratios is described in detail in Cochran (1977, Section 6.17) and Mosteller and Tukey (1977, Section 8C). The performances of the jackknife variance and (6.28) were compared in J. Rao and Beegle (1967), J. Rao (1969), and P. Rao and J. Rao (1971). The results on which one is superior are inconclusive. The two variance estimators seem to perform similarly except that the jackknife can be more erratic in small samples. The jackknife may tend to overestimate the variance of $\hat{\theta}$ and (6.28) to underestimate the variance. Clearly, the jackknife requires considerably more computation.

For unpaired data the effects of normality and nonnormality on the distribution of $\bar{y}/\bar{x}$ are analogous to the paired data case just discussed. The central limit theorem provides approximate normality for $\bar{y}/\bar{x}$ in moderate to large samples. In smaller samples there may be some positive skeweness. Outliers can be troublesome.

The pivotal statistic (6.13) and the test statistic (6.19) are two sample t statistics based on the assumption of equal variances for x and y. If this assumption is false, the effects are analogous to those described in Section 2.3.1 for the two sample problem.

When the evidence for σ_y^2 being substantially different from σ_x^2 is sufficiently strong, an appropriate reaction is to use Welch's t' statistic (see Section 2.3.3, "Other Tests"). The sample variances s_x^2 and s_y^2 are not pooled as in (6.12), and the test statistic

$$\frac{\bar{y} - \theta_0 \bar{x}}{\sqrt{\frac{s_y^2}{n} + \frac{\theta_0^2 s_x^2}{m}}} \tag{6.29}$$

is used in place of (6.19). The approximate degrees of freedom associated with (6.29) are

$$\hat{\nu} = \frac{\left(\frac{s_y^2}{n} + \frac{\theta_0^2 s_x^2}{m}\right)^2}{\frac{1}{n-1}\left(\frac{s_y^2}{n}\right)^2 + \frac{1}{m-1}\left(\frac{\theta_0^2 s_x^2}{m}\right)^2}. \tag{6.30}$$

Since the pivotal statistic for confidence intervals

$$\frac{\bar{y} - \theta \bar{x}}{\sqrt{\frac{s_y^2}{n} + \frac{\theta^2 s_x^2}{m}}} \tag{6.31}$$

has varying θ, the degrees of freedom associated with it should be the conservative lower bound $\min\{m-1, n-1\}$.

There have been no studies of the effects of dependent structures such as serial correlation on the ratio $\bar{y}/\bar{x}$ for paired or unpaired data. One is left to infer what one can from the one sample problem.

Exercises.

1. Use the delta method to justify the large sample standard deviation estimate for $\hat{\theta}$ employed in the confidence interval (6.10) [see also (6.28)].
2. Use the delta method to justify the approximation (6.17)–(6.18).
3. Show how

    (a) Beale's estimator (6.22),

(b) Tin's estimator (6.23),

(c) the jackknife estimator (6.26)

eliminate the $1/n$ bias term from $E(\bar{y}/\bar{x})$.

4. A study of structural evolutionary change utilized the specific adherence of lymphocytes to specialized lymphocyte-binding high endothelial venules (HEV) in lymph nodes.* When sample lymphocytes from other vertebrate species are perfused through mice, their exact number is unknown so the number of sample cells adhering to HEV must be scaled by the number of standard cells also adhering to HEV.

   The table gives 59 number pairs of sample (y) and standard (x) cells adhered to HEV.

   (a) Compute $\bar{y}/\bar{x}$.

   (b) Compute an estimated standard deviation for $\bar{y}/\bar{x}$.

   (c) Compute Beale's estimator (6.22).

   (d) Compute Tin's estimator (6.23).

   (e) Compute the jackknife estimator (6.26).

---

* Butcher, E., Scollay, R., and Weissman, I. (1979). Evidence of continuous evolutionary change in structures mediating adherence of lymphocytes to specialized venules. *Nature (London)* **280**, 496–498.

Standard	Sample	Standard	Sample
1	1	1	1
2	1	3	0
2	0	1	3
1	3	1	1
4	4	2	1
0	2	3	4
2	1	3	4
1	4	2	1
2	1	3	3
2	4	1	2
3	3	2	0
1	1	0	2
1	1	1	3
4	3	3	1
2	1	1	1
2	2	1	1
0	2	2	3
1	2	3	3
0	2	1	3
1	3	2	4
3	2	2	2
1	3	0	3
0	2	3	0
1	2	1	3
3	3	1	1
2	2	3	1
2	0	2	2
3	6	1	1
2	0	1	1
3	4		

5. In a study of diabetes, 21 patients, characaterized as normal, mild diabetic, and severe diabetic by a previous glucose tolerance test, were subjected to a constant glucose infusion.* Their steady-state values before and during the infusion and the increases Δ (= during − before) are given in the table.

   The investigator was intrested in whether there were any differences in the insulinogenic index $\Delta I/\Delta G$ between the 3 groups. What is your answer to this question?

---

* Reaven, G. and Miller, R. (1968). Study of the relationship between glucose and insulin responses to an oral glucose load in man. *Diabetes* **17**, 560–56.

Glucose Concentration (G)			Insulin Concentration (I)		
Before	During	ΔG	Before	During	ΔI
Patients With Normal Glucose Tolerance					
86	150	64	26	53	27
80	174	94	11	46	35
73	137	64	16	72	56
81	166	85	28	57	29
84	153	69	12	48	36
82	170	88	24	210	186
82	164	82	24	51	27
Patients With Mild Diabetes					
88	198	110	21	72	51
131	300	169	76	264	188
105	238	133	32	102	70
93	193	100	18	64	46
93	220	127	29	163	134
94	187	93	34	138	104
98	217	119	30	75	45
Patients With Severe Diabetes					
154	330	176	26	42	16
164	324	160	49	111	62
185	379	194	26	44	18
254	426	172	44	61	17
175	370	195	44	90	46
157	286	129	32	83	51
320	486	166	22	46	24

Chapter 7

VARIANCES

The previous chapters in this book have been predominately concerned with the estimation and testing of mean values. The only exceptions to this have been the estimation and testing of the variances of random effects in Chapters 3 and 4. This preoccupation with means is caused by most questions in applications being concerned with differences in location of different data sets. However, questions about variability do arise either as the primary issue as in random effects variance component problems and in deciding which of several measurements is more reproducible, or as a secondary issue as in deciding whether to pool sample variances.

This chapter focuses on inferences about variances for one, two, and more than two populations. The first section describes the statistical methods based on the assumption of an underlying normal distribution. Since all of these normal theory procedures are so very sensitive to departures from normality, the second section on nonnormality contains considerable discussion of alternative robust methods that are safer to use in applications.

7.1. Normal Theory.

Consider the one sample problem first. Let $y_1, \cdots, y_n$ be independently distributed as $N(\mu, \sigma^2)$.

To test the hypothesis $H_0 : \sigma^2 = \sigma_0^2$ against the two-sided alternative $H_1 : \sigma^2 \neq \sigma_0^2$, the ratio

$$\chi^2 = \frac{(n-1)s^2}{\sigma_0^2} \tag{7.1}$$

is used, where s^2 is the sample variance. Under H_0 this ratio has a χ^2 distribution with $n-1$ df so the test rejects when

$$s^2 < \frac{\sigma_0^2 \chi_{n-1}^{2\,1-(\alpha/2)}}{n-1} \quad \text{or} \quad \frac{\sigma_0^2 \chi_{n-1}^{2\,\alpha/2}}{n-1} < s^2, \tag{7.2}$$

where $\chi_{n-1}^{2\,1-(\alpha/2)}$ and $\chi_{n-1}^{2\,\alpha/2}$ are the lower and upper $100(\alpha/2)$ percentiles of a χ_{n-1}^2 distribution. The test (7.2) is called a χ^2 *test* and is essentially the likelihood ratio test. The latter uses slightly different critical constants. A two-sided P value is obtained by doubling the probability in the lower or upper tail of the χ_{n-1}^2 distribution beyond the observed value of $(n-1)s^2/\sigma_0^2$.

For a $100(1-\alpha)\%$ confidence interval the pivotal statistic s^2/σ^2 yields the interval

$$\frac{(n-1)s^2}{\chi_{n-1}^{2\,\alpha/2}} < \sigma^2 < \frac{(n-1)s^2}{\chi_{n-1}^{2\,1-(\alpha/2)}}. \tag{7.3}$$

A slightly shorter interval could be obtained by not restricting the probabilities in each tail to be equal (see Murdock and Williford, 1977).

For a one-sided test against $H_1 : \sigma^2 < \sigma_0^2$ or $H_1 : \sigma^2 > \sigma_0^2$ there is just one inequality in (7.2) with the whole significance level α being placed in one tail. Similarly, a one-sided P value is calculated form the single tail in the direction of the alternative. One-sided confidence intervals can also be obtained.

Problems of comparison of two or more variances arise more frequently than the one sample problem just discussed. For multiple samples, let $y_{i1}, \cdots, y_{in_i}$ be independently distributed as $N(\mu_i, \sigma_i^2)$, $i = 1, \cdots, I$, with independence between the samples for different i. The sample variances

$$s_i^2 = \frac{1}{n_i - 1} \sum_{j=1}^{n_i} (y_{ij} - \bar{y}_{i\cdot})^2 \tag{7.4}$$

are unbiased estimates of the corresponding population variances σ_i^2, $i = 1, \cdots, I$.

In the case of two samples ($I = 2$), the likelihood ratio of $H_0 : \sigma_1^2 = \sigma_2^2$ versus $H_1 : \sigma_1^2 \neq \sigma_2^2$ leads to the ratio

$$F = \frac{s_1^2}{s_2^2}, \tag{7.5}$$

which has an F distribution with $n_1 - 1$ and $n_2 - 1$ degrees of freedom under H_0. The two-sided *F test* would reject H_0 when (7.5) exceeds the upper $100(\alpha/2)$ percentile or falls below the lower $100(\alpha/2)$ percentile of the F distribution:

$$\frac{s_1^2}{s_2^2} < F_{n_1-1,n_2-1}^{1-(\alpha/2)} \quad \text{or} \quad \frac{s_1^2}{s_2^2} > F_{n_1-1,n_2-1}^{\alpha/2}. \tag{7.6}$$

The actual likelihood ratio test uses slightly different critical constants in (7.6). A two-sided P value is calculated by doubling the probability in the tail of the F_{n_1-1,n_2-1} distrbution beyond the observed s_1^2/s_2^2.

A $100(1-\alpha)\%$ confidence interval is constructed from the pivotal ratio $(s_1^2/\sigma_1^2)/(s_2^2/\sigma_2^2)$:

$$(1/F_{n_1-1,n_2-1}^{\alpha/2}) \left(\frac{s_1^2}{s_2^2} \right) < \frac{\sigma_1^2}{\sigma_2^2} < (1/F_{n_1-1,n_2-1}^{1-(\alpha/2)}) \left(\frac{s_1^2}{s_2^2} \right). \tag{7.7}$$

One-sided tests, P values, and confidence intervals can be computed.

The reader will have noticed by now that ratios of estimates and parameters are playing a key role in these variance problems whereas differences were central to mean problems. This is because questions about dispersion are ones of scale changes which lead to multiplicative factors.

For $I > 2$ samples there are three tests that share the limelight.

Bartlett's (1937) *test*, which is a slight modification of the likelihood ratio test, rejects the null hypothesis $H_0 : \sigma_1^2 = \cdots = \sigma_I^2$ when the statistic

$$M_1 = (N - I)\ln s_{\text{pool}}^2 - \sum_{i=1}^{I}(n_i - 1)\ln s_i^2 \tag{7.8}$$

exceeds the upper 100α percentile of the χ^2 distribution with $I-1$ df for large samples, where $N = \sum_{i=1}^{I} n_i$, "ln" is the natural logarithm, and s_{pool}^2 is the pooled sample variance, i.e.,

$$s_{\text{pool}}^2 = \frac{1}{N - I}\sum_{i=1}^{I}(n_i - 1)s_i^2. \tag{7.9}$$

For smaller samples M_1 is approximately distributed as

$$(1 + A)\chi_{I-1}^2, \tag{7.10}$$

where

$$A = \frac{1}{3(I - 1)}\left[\left(\sum_{i=1}^{I}\frac{1}{n_i - 1}\right) - \frac{1}{N - I}\right]. \tag{7.11}$$

For very small n_i tables are given in Pearson and Hartley (1970).

The next two tests are not as general as Bartlett's test in that the sample sizes need to be equal (i.e., $n_i \equiv n$).

Hartley's (1950) *test* compares the statistic

$$M_2 = \frac{s_{\max}^2}{s_{\min}^2}, \tag{7.12}$$

where

$$s^2_{\max} = \max\{s^2_1, \cdots, s^2_I\},$$
$$s^2_{\min} = \min\{s^2_1, \cdots, s^2_I\}, \tag{7.13}$$

with the upper 100α percentile for the distribution of this ratio under H_0. Tables of this maximum F ratio were given by David (1952) for $\alpha = .05, .01$, $I = 2(1)12$, and $n - 1 = 2(1)10$, 12, 15, 20, 30, 60, ∞. These tables are reproduced in Owen (1962) and Pearson and Hartley (1970).

Cochran's (1941) *test* compares the statistic

$$M_3 = \frac{s^2_{\max}}{\sum_{i=1}^{I} s_i^2} \tag{7.14}$$

with the upper 100α percentile for the distribution of this ratio under H_0. Tables are given in Eisenhart and Solomon (1947) for $\alpha = .05$, .01, $I = 2(1)12$, 15, 20, 24, 30, 40, 60, 120, ∞, and $n - 1 = 1(1)10$, 16, 36, 144, ∞, and these are reproduced in Pearson and Hartley (1970) for I up to 20.

The statistics M_1 and M_3 do not lend themselves to the development of multiple comparisons procedures, but M_2 does. In particular, with probability $1 - \alpha$

$$\frac{1}{M_2^\alpha}\left(\frac{s_i^2}{s_{i'}^2}\right) \le \frac{\sigma_i^2}{\sigma_{i'}^2} \le M_2^\alpha\left(\frac{s_i^2}{s_{i'}^2}\right) \quad \text{for all} \quad i, i', \tag{7.15}$$

where M_2^α is the upper 100α percentile of the M_2 distribution for I populations with $n - 1$ df for each s_i^2.

No techniques have appeared in the literature that are tailored to testing the null hypothesis $H_0 : \sigma_1^2 = \cdots = \sigma_I^2$ against the ordered alternative $H_1 : \sigma_1^2 \le \sigma_2^2 \le \cdots \le \sigma_I^2$ with strict inequality at least once.

7.2. Nonnormality.

7.2.1. Effect

The effects of nonnormality on the distribution theories for the test statistics (7.1), (7.5), (7.8), (7.12), and (7.14) are catastrophic. For each test the actual significance level can be considerably different from the nominally stated level. For a heavy-tailed distribution the probability of rejection under H_0 greatly exceeds α, and for a short-tailed distribution the probablity is considerably less than α.

Pearson (1931) first pointed out this sensitivity in the two sample problem through the use of sampling experiments. These results were later confirmed theoretically by Geary (1947), Finch (1950), and Gayen (1950a). Box (1953) found the effects to be even more extreme with three or more populations. Pearson and Please (1975) carried out extensive simulations for one and two samples.

To given an indication of the magnitude of the effect, Geary (1947) in his Table 1 gives the probability .166 of rejecting H_0 in large samples with an $\alpha = .05$ F test when H_0 is true but the underlying distribution has kurtosis $\gamma_2 = 2$. Box (1953) in his Table 1 shows that for the M_1 test this increases to .315 for $I = 5$ and then to .489 for $I = 10$. On the other hand, for $\gamma_2 = -1$ Geary and Box give .0056, .0008, and .0001 for $I = 2$, 5, and 10, respectively, as the actual significance levels for a 5% level test with large samples.

The effect of skewness (i.e., $\gamma_1 \neq 0$) on the actual significance levels of the variance tests is much less extreme. Some values are given in Table 1 of Finch (1950), and numerous figures are given in Pearson and Please (1975).

The reason for this hypersensitivity can be seen in the variance of a single sample variance s^2. If the observations $y_1, \cdots, y_n$ entering $s^2 = \sum_{i=1}^{n}(y_i - \bar{y})^2/(n-1)$ are independently distributed according to a general distribution $F(y)$, then

$$E(s^2) = \sigma^2,$$
$$\operatorname{Var}(s^2) = \sigma^4 \left(\frac{2}{n-1} + \frac{\gamma_2}{n} \right), \tag{7.16}$$

where $\gamma_2 = \gamma_2(y)$ is the kurtosis of $F(y)$. The distribution theory for the test statistic (7.1) is based on the normality of the y_i, which implies that the variance of s^2 is $2\sigma^4/(n-1)$ [i.e., (7.16) with $\gamma_2 = 0$]. Nothing informs the critical points in (7.2) that the variability of s^2 is larger than this when $\gamma_2 > 0$ and is smaller than this when $\gamma_2 < 0$.

Similar phenomena occur for the other test statistics (7.5), (7.8), (7.12), and (7.14). Their variability is greater or less than that presupposed by normal theory depending on whether $\gamma_2 > 0$ or $\gamma_2 < 0$.

This situation is very different from the t tests of Chapters 1 and 2. There the standard deviation of the numerator statistic is correctly estimated by the denominator regardless of whether the data are normally distributed.

The ANOVA F tests of Chapters 3 and 4 for location differences do not have the same sensitivity to nonnormality as the F test based on (7.5) because the numerator mean sum of squares is computed from mean values in which the kurtosis is diminished [see (1.10)]. The denominator mean sum of squares scales the statistic to have the correct approximate expectation. Also, the denominator usually has sufficiently large degrees of freedom that its variability is not an important factor.

In short, F tests for location are reasonably robust, but F tests for dispersion are not.

7.2.2. Detection

The problem of detecting nonnormality when the inference is concerned with variances is no different than in the one sample, two sample, and one-way classification designs for location inference. Probit plots of the data in each sample are especially recommended, and tests of normality are also available. The reader is referred to Section 1.2.2 for a full discussion of these graphical and testing procedures.

7.2.3. Correction

There is an abundance of *nonparametric rank tests* for the two sample dispersion problem. Unfortunately, none of them is much good for applications.

Perhaps the best known is a test proposed by Freund and Ansari (1957), Barton and David (1958), and Ansari and Bradley (1960). It assigns rank 1 to the smallest and largest observations in the combined data set from the two samples, rank 2 to the second smallest and second largest, etc. The test statistic is the sum of the ranks associated with the observations in one of the samples. Small sample tables and large sample means and variances are available. For greater detail and discussion the reader should consult Hollander and Wolfe (1973) or Gibbons (1971).

For this test to not give misleading results the population medians must be equal. Moses (1963) gives examples of what can happen when the medians are not equal. Since medians between two populations are hardly ever known to be equal in applications, the worth of this procedure is in question. Moreover, there is no escape from this judgment by subtracting the sample median from the data in each sample, i.e., by ranking $y_{1j} - m_1$, $j = 1, \cdots, n_1$, combined with $y_{2j} - m_2$, $j = 1, \cdots, n_2$, where m_1 and m_2 are the medians of $\{y_{1j}\}$ and $\{y_{2j}\}$, respectively. Standardization by median subtraction produces a test which is not distribution-free.

S. Siegel and Tukey (1960) proposed a test with a different ranking scheme from the aforementioned one, but it is essentially equivalent to the Ansari-Barton-Bradley-David-Freund test (see Gibbons,

1971, Hájek and Šidák, 1967, or Klotz, 1962).

There have been a number of other rank tests based on squared central ranks, normal scores, etc. For a complete list of references see Hollander and Wolfe (1973, Chapter 5). However, all these tests require some assumption concerning known or equal medians, and these assumptions cannot be relaxed while still preserving a distribution-free test. In addition, it can be argued that rank tests do not make sense without a restriction on the locations because a monotonic transformation, which preserves the ranks, can create unequal sizes of variation between two populations with identical dispersions but unequal locations. For an example and discussion see Moses (1963).

Box (1953) proposed a procedure based on *grouped data*. Select a group size k, and divide each sample n_i into g_i groups of size k. Hopefully, k can be selected so that $g_i \cdot k$, which has to be smaller than n_i, is very close to n_i for all i because the remaining $n_i - (g_i \cdot k)$ observations are discarded in each sample. For the k observations in the jth group of the ith sample, let s_{ij}^2 be their sample variance. The s_{ij}^2 are identically distributed within each population, and they are all independent. Define

$$z_{ij} = \log s_{ij}^2, \quad i = 1, \cdots, I, \quad j = 1, \cdots, g_i. \tag{7.17}$$

Since

$$E(z_{ij}) \cong \log \sigma_i^2, \tag{7.18}$$

with the approximation improving as k increases, Box's proposal is to treat the z_{ij} as observations in a location problem in order to test hypotheses about the σ_i^2, $i = 1, \cdots, I$, or construct confidence intervals. For $I = 1$ and 2, t tests and confidence intervals can be applied, and for $I > 2$ the techniques of analysis of variance, multiple comparisons, and monotone alternatives are appropriate.

The log transformation is applied to s_{ij}^2 in (7.17) in order to make the distribution of z_{ij} more symmetric. However, without it $E(z_{ij})$ would exactly equal σ_i^2. Thus there is a trade-off in the use or nonuse of log in terms of whether symmetrization is more important than an exact expectation or vice versa.

A major question is how to select k. As k is increased, the symmetry of the distribution of s_{ij}^2 or $\log s_{ij}^2$ is improved and (7.18) becomes more exact, but the number of groups g_i in each sample decreases so the t and ANOVA analyses lose power and the confidence intervals become broader. Shorack (1969) recommends selecting k as large as possible but not exceeding 10 while preserving reasonable sizes for $g_1, \cdots, g_I$. I would suggest having k at least 5 if at all possible, and then seeing what increasing k does to the g_i.

Clearly, this method throws away information. Some observations may be discarded, and no comparisons are made between the y_{ij} in different groups within a sample. Also, different groupings of the data within each sample have the potential to produce substantially different answers. Nonetheless, simulation studies show that this technique works satisfactorily in an inefficient manner. If one can afford to be inefficient because of an overabundance of data, this technique is easy to apply and interpret.

Moses (1963) suggested applying nonparametric rank tests to the Box s_{ij}^2, $i = 1, \cdots, I$, $j = 1, \cdots, g_i$. For example, run the Wilcoxon rank test on the $g_1 + g_2$ values of s_{ij}^2 in the two sample problem. Shorack (1966) extended this idea to obtaining point and interval estimates. The power of this test in the two sample problem is compared with other competitors in Miller (1968) and Shorack (1969). Like the original Box test, it is reliable but inefficient.

There are three approximate *robust tests* that do not have the unrealistic assumptions of the nonparametric rank tests applied to the y_{ij} and utilize the data in an ungrouped fashion.

The first is a simple idea due to Levene (1960). The *Levene s test* is to treat the values $z_{ij} = (y_{ij} - \bar{y}_{i\cdot})^2$, $j = 1, \cdots, n_i$ in each sample i, $i = 1, \cdots, I$, as though they are independently, identically, normally distibuted under H_0, and to apply the usual t and ANOVA tests and confidence intervals to them.

Clearly, the z_{ij} do not satisfy the assumptions imposed on them. Within a sample they are not independent because of the common $\bar{y}_{i\cdot}$, but the correlation is of order $1/n_i^2$. They are not identically distributed under H_0 unless $n_i \equiv n$, but any departure from this has a minor effect. They are not normally distributed, but the ANOVA procedures for location inference are reasonably robust for nonnormality.

In spite of worries over the assumptions, the Monte Carlo studies reported in Levene (1960), Miller (1968), Shorack (1969), and M. B. Brown and Forsythe (1974), demonstrate that the Levene *s* test performs quite satisfactorily. It has reasonably good robustness for validity against nonnormal distributions. However, its power against heavy-tailed alternatives is not quite as good as that of the Box-Andersen and jackknife tests described next.

Levene (1960) also proposed applying the preceding idea to $z_{ij} = |y_{ij} - \bar{y}_{i\cdot}|$, $z_{ij} = \log|y_{ij} - \bar{y}_{i\cdot}|$, and $z_{ij} = |y_{ij} - \bar{y}_{i\cdot}|^{1/2}$, but the test with $z_{ij} = (y_{ij} - \bar{y}_{i\cdot})^2$ is the generally accepted version. A small difficulty with using $z_{ij} = |y_{ij} - \bar{y}_{i\cdot}|$ is pointed out in Miller (1968). M. B. Brown and Forsythe (1974) consider variations on the Levene approach with different location estimates.

Box and Andersen (1955) applied permutation theory to construct approximate robust tests. To understand their procedure, consider the hypothetical two sample dispersion problem with known population means, which for simplicity are assumed to have been subtracted already from the observations.

The moments of the test statistic are considered under two different distribution theories. Normal theory assumes that the data are normally distributed. Permutation theory assumes that the two samples have been randomly selected without replacement from $u_1, \cdots, u_{n_1+n_2}$ where

$$u_1 = y_{11}, \cdots, u_{n_1} = y_{1n_1}, u_{n_1+1} = y_{21}, \cdots, u_{n_1+n_2} = y_{2n_2}. \tag{7.19}$$

Each of the $\binom{n_1+n_2}{n_1}$ possible combinations is equally likely.

Rather than computing the moments of the $F = s_1^2/s_2^2$ statistic, it is simpler to calculate the moments of the related ratio

$$B = \frac{\sum_{j=1}^{n_1} y_{1j}^2}{\sum_{j=1}^{n_1} y_{1j}^2 + \sum_{j=1}^{n_2} y_{2j}^2} \tag{7.20}$$

because the denominator remains constant for the permutation distribution. To reject for small or large B is equivalent to rejecting for small or large F.

The theoretical mean of B is the same whether it is computed under the assumption of a normal distribution or under the permutation distribution:

$$E_N(B) = E_P(B) = \frac{n_1}{N}, \tag{7.21}$$

where $N = n_1 + n_2$. However, the theoretical variances differ. Under the normal distribution

$$\mathrm{Var}_N(B) = \frac{2n_1 n_2}{N^2(N+2)}, \tag{7.22}$$

and under the permutation distribution

$$\mathrm{Var}_P(B) = \frac{2n_1 n_2}{N^2(N+2)}\left[1 + \frac{1}{2}\left(\frac{N}{N-1}\right)(b_2 - 3)\right], \tag{7.23}$$

where

$$b_2 = \frac{(N+2)\sum_{i=1}^{N} u_i^4}{\left(\sum_{i=1}^{N} u_i^2\right)^2}. \tag{7.24}$$

The variance (7.23) can be made to equal the variance (7.22) if new sample sizes n_1^* and n_2^* are used in (7.22) where $n_1^* = dn_1$, $n_2^* = dn_2$, and

$$d = \left[1 + \frac{1}{2}\left(\frac{N+2}{N+2-b_2}\right)(b_2 - 3)\right]^{-1}. \tag{7.25}$$

The mean of B remains unchanged under this substitution since $n_1^*/(n_1^* + n_2^*) = n_1/N$. Thus the normal theory distribution for B can be made to approximate the permutation theory distribution for B by redefining the sample sizes as described.

This suggests the following approximate *Box-Andersen test*. It is called the *APF-test* by Shorack (1969). Calculate the usual F statistic (7.5), but compare it with the critical points of an F distribution on

$$\hat{d}(n_1 - 1) \quad \text{and} \quad \hat{d}(n_2 - 1) \tag{7.26}$$

degrees of freedom where

$$\hat{d} = \left[1 + \frac{1}{2}(\hat{b}_2 - 3)\right]^{-1} \tag{7.27}$$

and

$$\hat{b}_2 = \frac{\left[\sum_{i=1}^2 n_i\right]\left[\sum_{i=1}^2 \sum_{j=1}^{n_i} (y_{ij} - \bar{y}_{i\cdot})^4\right]}{\left[\sum_{i=1}^2 \sum_{j=1}^{n_i} (y_{ij} - \bar{y}_{i\cdot})^2\right]^2}. \tag{7.28}$$

Because of the closeness of the first two moments, the normal theory critical points are good approximations to those that would have been obtained for the permutation distribution.

The Monte Carlo studies in Box and Andersen (1955), Miller (1968), and Shorack (1969) demonstrate that the Box-Andersen test maintains the correct approximate significance level under the null hypothesis for a variety of heavy and short-tailed distributions and has superior power to the other competitive tests with the exception of the jackknife, which performs approximately the same.

Box and Andersen also considered the $I > 2$ sample problem. For I samples the analogous procedure is to compare $\hat{d}M_1$ with critical points from a χ^2_{I-1} distribution, where $\hat{d}$ is given by (7.27) and $\hat{b}_2$ in(7.28) is calculated by summing over all I samples in the numerator and denominator.

Layard (1973) proposed a somewhat different statistic for the $I > 2$ problem which also involves an estimate of the population kurtosis.

There is no possibility of a permutation distribution in the one sample problem, but a different approach yields an analogous procedure. From (7.16)

$$\mathrm{Var}(s^2) \cong \frac{2\sigma^4}{n-1}\left(1 + \frac{1}{2}\gamma_2\right) \tag{7.29}$$

for any underlying distribution. Since the variance of s^2 is exactly $2\sigma^4/(n-1)$ under normal theory, the variance of s^2 for an arbitrary distribution is approximately equal to the normal theory variance with degrees of freedom $\hat{d}(n-1)$ where

$$\hat{d} = \left(1 + \frac{1}{2}\hat{\gamma}_2\right)^{-1} \tag{7.30}$$

and the sample estimate of the kurtosis is

$$\hat{\gamma}_2 = \frac{n\sum_{j=1}^{n}(y_j - \bar{y})^4}{\left[\sum_{j=1}^{n}(y_j - \bar{y})^2\right]^2} - 3. \tag{7.31}$$

Thus $\hat{d}(n-1)s^2/\sigma^2$ is approximately distributed as a χ^2 variable with $\hat{d}(n-1)$ degrees of freedom. Tests or confidence intervals for σ^2 can be constructed from this pivotal statistic.

The *jackknife* is the final procedure to be mentioned. It is a general technique that has already been suggested in this book for variance component problems (Sections 3.6.3, 4.6, and 4.7.2), the correlation coefficient (Section 5.3.3), the errors-in-variables model (Section 5.7), and ratios (Section 6.2). A general review of the jackknife is given by Miller (1974a). Its specific application to the one sample variance problem is described in Mosteller and Tukey (1968, 1977), to the two sample problem in Miller (1968), and to $I > 2$ populations in Layard (1973).

The sample variances could be jackknifed directly, but jackknifing their logarithms produces better results. Thus for population i let $\theta_i = \log \sigma_i^2$ and $\hat{\theta}_i = \log s_i^2$. The pseudo-values for $j = 1, \cdots, n_i$ are defined by

$$\tilde{\theta}_{ij} = n_i \hat{\theta}_i - (n_i - 1)\hat{\theta}_{i,-j}, \tag{7.32}$$

where the estimate $\hat{\theta}_{i,-j} = \log s_{i,-j}^2$ has the jth observation in sample i deleted. The $\tilde{\theta}_{ij}$ should be treated as independent observations (even though they are not), which are identically distributed in sample i with approximate mean θ_i. For the one and two sample problems t statistics can be computed from the $\tilde{\theta}_{ij}$ to test hypotheses and construct confidence intervals for θ_1 and θ_1, θ_2, respectively (see Chapters 1 and 2). Taking antilogs of the endpoints transforms any confidence interval back into a confidence interval in the original variance scale. For $I > 2$ populations a one-way ANOVA can be used for testing $H_0 : \theta_1 = \cdots = \theta_I$ (i.e., $H_0 : \sigma_1^2 = \cdots = \sigma_I^2$). Multiple comparisons and monotone alterntives techniques are also available (see Chapter 3, "Fixed Effects").

The simulations of M. B. Brown and Forsythe (1974) suggest that for very unequal sample sizes in the two sample problem (e.g., $n_2 = 2n_1$ with $n_1 = 10$, 20) the actual significance levels for the jackknife exceed the nominal levels when the distribution is heavy-tailed. This may be due to pooling somewhat unequal variances calculated from pseudo-values. Given a choice one would prefer a balanced experiment. However, if this is not the case, then the variation in the pseudo-values should be examined to determine the best method for their statistical analysis.

The logarithmic transformation seems natural to use with the sample variance because it keeps the variance estimate associated with each pseudo-value (i.e., antilog $\tilde{\theta}_{ij}$) positive and it is the variance stabilizing transformation. Cressie (1981) gives a theoretical foundation for this selection, and empirical work supports its use.

The Monte Carlo studies in Miller (1968) and Layard (1973) establish the jackknife as a robust procedure for testing equality of variances that has power equivalent to the Box-Andersen test. Each test in its own way is using the data to estimate the fourth central moment of the underlying distribution. This moment controls the variation in s^2. Levene's s test is also doing the same thing but not quite as effectively.

Since the jackknife and Box-Andersen tests are essentially equivalent in performance, which one should be chosen in practice? The choice may be made on computational grounds. Since neither is part of the standard computer packages, it may be a question of which is easier to program and implement.

Even though the jackknife and Box-Andersen tests are robust for heavy-tailed distributions, they are not resistant to outliers. Single or multiple aberrant values can grossly distort s^2 because the deviation is squared. The impact of an outlier is greater on s^2 than on $\bar{y}$. The jackknife and Box-Andersen tests can be misleading if one or more s_i^2 have been affected by outliers. Trimming the outlier(s) is the only known recourse, but the effect of this on the performance of the tests has not been studied.

To counteract the effects of outliers, one might consider alternative measures of dispersion that are more resistant to outliers. Unfortunately, this area is not well developed. Huber (1981) discusses L, M, and R-estimators of scale. My discussion is limited to brief descriptions of two estimators whose values are invariant under changes in the most extreme observations and therefore are resistant to outliers.

For a single sample $y_1, \cdots, y_n$ the *median absolute deviation (MAD)* is defined by

$$\text{MAD} = \text{median}\{|y_i - m|\}, \tag{7.33}$$

where $m = \text{median}\{y_i\}$. In words, MAD is the median of the absolute deviations of the observations from their median. It estimates the corresponding quantity defined in terms of population values. To date, the application of MAD seems to be limited to providing a scale estimate for use with M- estimators of location (see Section 1.2.3, "Robust Estimation").

The *interquartile range (IQR)* is the seventy-fifth percentile of the sample minus the twenty-fifth percentile. It estimates the corresponding percentile difference in the population. For a normal distribution the population interquartile range is related to the standard deviation by

$$\text{IQR} = 1.35\sigma \tag{7.34}$$

(to two decimal accuracy) and to the population MAD by

$$\text{IQR} = 2\text{MAD}. \tag{7.35}$$

Both MAD and IQR have seen limited use in applications. Neither has been investigated as a tool for testing the equality of dispersion between populations.

7.3. Dependence.

When there is known blocking in the data due to observations being taken at different times, with different equipment, etc., this must be taken into account in the statistical analysis. Variances need to be calculated within blocks. If there are enough blocks, this may be a blessing in disguise because then a Box type approach using the variation between block variances forms a basis for inference.

Little is known about the effects on variance tests for other types of dependence between the observations and how to properly correct for them. For example, serial dependence in the data changes the expectation [see (1.48)] and variance of s^2, but appropriate modifications to the procedures mentioned in this chapter have not been worked out.

Exercises.

1. Derive the normal theory likelihood ratio statistic for testing $H_0 : \sigma_i^2 \equiv \sigma^2$, $i = 1, \cdots, I$, versus $H_1 : \sigma_i^2 \not\equiv \sigma^2$. How is this statistic similar to Bartlett's M_1 statistic (7.8)?

2. Show that for $y_1, \cdots, y_n$ independently, identically distributed,

$$\text{Var}(s^2) = \sigma^4 \left(\frac{2}{n-1} + \frac{\gamma_2}{n} \right),$$

where s^2 is the sample variance, σ^2 is the population variance, and γ_2 is the population kurtosis.

3. Show that for $y_1, \cdots, y_n$ independently, identically distributed the correlation between $(y_i - \bar{y})^2$ and $(y_{i'} - \bar{y})^2$, $i \neq i'$, is of the order $1/n^2$.

4. For the data in Exercise 5 of Chapter 2, test the equality of variances between the R and NR groups by

   (a) the F test,

   (b) the Box-Andersen test,

   (c) the jackknife test.

5. For the data in Exercise 10 of Chapter 3, test the equality of variances between the three groups by

(a) Bartlett's M_1 test,

(b) the Box-Andersen test,

(c) the jackknife test.

6. Thermodilution cardiac output (TDCO) measurements are commonly employed as a useful adjunct in the management of critically ill patients. TDCOs often exhibit considerable variation in clinical settings. Prior to this study there had been no clinical studies revealing when to perform TDCOs in relation to the respiratory cycle.* In this Stanford study 32 patients were prospectively studied to compare TDCOs measured at peak-inspiration, at end-exhalation, and at random times in spontaneously breathing or mechanically ventilated patients. Three TDCO measurements were obtained at each of the 3 different times in the respiratory cycle in each patient. The data for 12 spontaneously breathing patients are displayed in the table.

   Determine if any of the three times show significant more (or less) variation in their 3 values than the others.

* Stevens, J. H., Raffin, T.A., Mihm, F. G., Rosenthal, M. H., and Stetz, C. W. (1985). Increased reproducibility of thermodilution cardiac output measurements in clinical practice. *Journal of the American Medical Association*, in press.

Patient	End-Exhalation			Peak-Inspiration			Random		
01	6.94	6.57	6.38	6.35	6.73	6.43	7.27	6.35	5.95
02	5.44	5.24	5.05	5.33	5.30	5.15	5.01	5.17	5.34
03	7.18	6.85	6.80	7.34	6.71	6.94	7.14	6.81	7.18
04	7.26	8.37	8.38	8.25	8.06	8.83	10.21	8.40	10.11
05	5.24	5.09	4.94	4.75	5.41	5.32	5.32	5.82	4.33
06	8.45	9.00	8.39	8.86	7.98	8.03	8.94	8.22	9.80
07	3.28	3.44	3.51	2.95	3.24	3.48	3.04	4.10	3.31
08	5.39	5.21	5.14	5.13	4.93	5.33	5.13	4.19	5.17
09	2.71	2.68	2.84	2.51	2.54	2.69	2.51	2.51	2.55
10	11.10	10.71	10.41	10.74	10.38	10.64	11.92	11.02	8.31
11	6.14	6.23	5.91	6.03	6.08	5.97	5.70	6.94	6.51
12	13.06	12.76	13.65	12.12	13.42	13.51	14.87	14.24	10.66

REFERENCES

Abelson, R. P. and Tukey, J. W. (1963). Efficient utilization of non-numerical information in quantitative analysis: general theory and the case of simple order. *Annals of Mathematical Statistics* **34**, 1347–1369.

Abraham, J. K. (1960). Note 154: On an alternative method of computing Tukey's statistic for the Latin square model. *Biometrics* **16**, 686–691.

Amemiya, T. (1973). Regression analysis when the variance of the dependent variable is proportional to the square of its expectation. *Journal of the American Statistical Association* **68**, 928–934.

Andersen, A. H., Jensen, E. B., and Schou, G. (1981). Two-way analysis of variance with correlated errors. *International Statistical Review* **49**, 153–167.

Anderson, R. L. (1942). Distribution of the serial correlation coefficient. *Annals of Mathematical Statistics* **13**, 1–13.

Anderson, R. L. and Bancroft, T. A. (1952). *Statistical Theory in Research.* McGraw-Hill, New York.

Anderson, T. W. (1958). *An Introduction to Multivariate Statistical Analysis.* Wiley, New York.

Anderson, T. W. (1971). *The Statistical Analysis of Time Series.* Wiley, New York.

Anderson, T. W. (1976). Estimation of linear functional relationships: approximate distributions and connections with simultaneous equations in econometrics. *Journal of the Royal Statistical Society, Series B* **38**, 1–20.

Andrews, D. F. (1971a). Significance tests based on residuals. *Biometrika* **58**, 139–148.

Andrews, D. F. (1971b). A note on the selection of data transformations. *Biometrika* **58**, 249–254.

Andrews, D. F. (1974). A robust method for multiple linear regression. *Technometrics* **16**, 523–531.

Andrews, D. F., Bickel, P. J., Hampel, F. R., Huber, P. J., Rogers, W. H., and Tukey, J. W. (1972). *Robust Estimation of Location: Survey and Advances.* Princeton University Press, Princeton, N.J.

Ansari, A. R. and Bradley, R. A. (1960). Rank-sum tests for dispersions. *Annals of Mathematical Statistics* **31**, 1174–1189.

Anscombe, F. J. (1961). Examination of residuals. *Proceedings of the Fourth Berkeley Symposium on Mathematical Statistics and Probability*, Vol. 1, (J. Neyman, Ed.). University of California Press, Berkeley, pp. 1–52.

Armitage, P. (1955). Tests for linear trends in proportions and frequencies. *Biometrics* **11**, 375–386.

Armitage, P. (1971). *Statistical Methods in Medical Research.* Blackwell, Oxford.

Arvesen, J. N. (1969). Jackknifing U-statistics. *Annals of Mathematical Statistics* **40**, 2076–2100.

Arvesen, J. N. and Layard, M. W. J. (1975). Asymptotically robust tests in unbalanced variance component models. *Annals of Statistics* **3**, 1122–1134.

Arvesen, J. N. and Schmitz, T. H. (1970). Robust procedures for variance component problems using the jackknife. *Biometrics* **26**, 677–686.

Atkinson, A. C. (1973). Testing transformations to normality. *Journal of the Royal Statistical Society, Series B* **35**, 473–479.

Barlow, R. E., Bartholomew, D. J., Bremner, J. M., and Brunk, H. D. (1972). *Statistical Inference under Order Restrictions.* Wiley, New York.

Barnett, V. and Lewis, T. (1978). *Outliers in Statistical Data.* Wiley, New York.

Bartholomew, D. J. (1959a). A test of homogeneity for ordered alternatives. *Biometrika* **46**, 36–48.

Bartholomew, D. J. (1959b). A test of homogeneity for ordered alternatives. II. *Biometrika* **46**, 328–335.

Bartholomew, D. J. (1961a). A test of homogeneity of means under restricted alternatives. *Journal of the Royal Statistical Society, Series B* **23**, 239–272.

Bartholomew, D. J. (1961b). Ordered tests in the analysis of variance. *Biometrika* **48**, 325–332.

Bartlett, M. S. (1937). Properties of sufficiency and statistical tests. *Proceedings of the Royal Society (London), Series A* **160**, 268–282.

Bartlett, M. S. (1949). Fitting a straight line when both variables are subject to error. *Biometrics* **5**, 207–212.

Barton,D. E. and David, F. N. (1958). A test for birth order effect. *Annals of Human Genetics* **22**, 250–257.

Beale, E. M. L. (1962). Some uses of computers in operational research. *Industrielle Organisation* **31**, 51–52.

Bement, T. R. and Williams, J. S. (1969). Variance of weighted regression estimators when sampling errors are independent and heteroscedastic. *Journal of the American Statistical Association* **64**, 1369–1382.

Benard, A. and van Elteren,P. (1953). A generalization of the method of m rankings. *Indagationes Mathematicae* **15**, 358–369.

Berkson, J. (1950). Are there two regressions? *Journal of the American Statistical Association* **45**, 164–180.

Berkson, J. (1955). Maximum likelihood and minimum χ^2 estimates of the logistic function. *Journal of the American Statistical Association* **50**, 130–162.

Bickel, P. J. (1973). On some analogues to linear combinations of order statistics in the linear model. *Annals of Statistics* **1**, 597–616.

Bickel, P. J. (1978). Using residuals robustly I: tests for heteroscedasticity, nonlinearity. *Annals of Statistics* **6**, 266–291.

Bickel, P. J. and Doksum, K. A. (1981). An analysis of transformations revisited. *Journal of the American Statistical Association* **76**, 296–311.

Birch, M. W. (1964). A note on the maximum likelihood estimation of a linear structural relationship. *Journal of the American Statistical Association* **59**, 1175–1178.

Bishop, Y. M. M., Fienberg, S. E., and Holland,P. W. (1975). *Discrete Multivariate Analysis: Theory and Practice.* MIT Press, Cambridge.

Box, G. E. P. (1953). Non-normality and tests on variances. *Biometrika* **40**, 318–335.

Box, G. E. P. (1954a). Some theorems on quadratic forms applied in the study of analysis of variance problems, I. Effect of inequality of variance in the one-way classification. *Annals of Mathematical Statistics* **25**, 290–302.

Box, G. E. P. (1954b). Some theorems on quadratic forms applied in the study of analysis of variance problems, II. Effect of inequality of variance and of correlation of errors in the two-way classification. *Annals of Mathematical Statistics* **25**, 484–498.

Box, G. E. P. and Andersen, S. L. (1955). Permutation theory in the derivation of robust criteria and the study of departures from assumption. *Journal of the Royal Statistical Society, Series B* **17**, 1–26.

Box, G. E. P. and Cox, D. R. (1964). An analysis of transformations. *Journal of the Royal Statistical Society, Series B* **26**, 211–243.

Box, G. E. P. and Cox, D. R. (1982). An analysis of transformations revisited, rebutted. *Journal of the American Statistical Association* **77**, 209–210.

Box, G. E. P. and Draper, N. R. (1975). Robust designs. *Biometrika* **62**, 347–352.

Box, G. E. P. and Hill, W. J. (1974). Correcting inhomogeneity of variance with power transformation weighting. *Technometrics* **16**, 385–389.

Brillinger, D. R. (1966). Discussion on Mr. Sprent's paper. *Journal of the Royal Statistical Society, Series B* **28**, 294.

Brown, G. W. and Mood, A. M. (1948). Questions and answers: homogeneity of several samples. *American Statistician* **2**(3), 22.

Brown, G. W. and Mood, A. M. (1951). On median tests for linear hypotheses. *Proceedings of the Second Berkeley Symposium on Mathematical Statistics and Probability*, (J. Neyman, Ed.). University of California Press, Berkeley, pp. 159–166.

Brown, M. B. and Forsythe, A. B. (1974). Robust tests for the equalty of variances. *Journal of the American Statistical Association* **69**, 364–367.

Brown, M. L. (1982). Robust line estimation with errors in both variables. *Journal of the American Statistical Association* **77**, 71–79.

Brown, R. A. (1974). Robustness of the studentized range statistic. *Biometrika* **61**, 171–175.

Brunden, M. N. and Mohberg, N. R. (1976). The Benard-van Elteren statistic and nonparametric computation. *Communications in Statistics, Part B* **5**,155–162.

Brunk, H. D. (1955). Maximum likelihood estimates of monotone parameters. *Annals of Mathematical Statistics* **26**, 607–616.

Brunk, H. D. (1958). On the estimation of parameters restricted by inequalities. *Annals of Mathematical Statistics* **29**, 437–454.

Bulmer, M. G. (1957). Approximate confidence limits for components of variance. *Biometrika* **44**, 159–167.

Carroll, R. J. (1980). A robust method for testing transformations to achieve normality. *Journal of the Royal Statistical Society, Series B* **42**, 71–78.

Carroll, R. J. and Ruppert, D. (1981a). On robust tests for heteroscedasticity. *Annals of Statistics* **9**, 206–210.

Carroll, R. J. and Ruppert, D. (1981b). On prediction and the power transformation family. *Biometrika* **68**, 609–615.

Carroll, R. J. and Ruppert, D. (1982). Robust estimation in heteroscedastic linear models. *Annals of Statistics* **10**, 429–441.

Chacko, V. J. (1963). Testing homogeneity against ordered alternatives. *Annals of Mathematical Statistics* **34**, 945–956.

Chan, N. N. and Mak, T. K. (1983). Estimation of multivariate linear functional relationships. *Biometrika* **70**, 263–267.

Chernoff, H. and Lieberman, G. J. (1954). Use of normal probability paper. *Journal of the American Statistical Association* **49**, 778–785.

Cochran, W. G. (1941). The distribution of the largest of a set of estimated variances as a fraction of their total. *Annals of Eugenics* (superseded by *Annals of Human Genetics*) **11**, 47–52.

Cochran, W. G. (1954). Some methods for strengthening the common χ^2 tests. *Biometrics* **10**, 417–451.

Cochran, W. G. (1977). *Sampling Techniques*, Third Edition. Wiley, New York.

Cochran, W. G. and Cox, G. M. (1957). *Experimental Designs*, Second Edition. Wiley, New York.

Conover, W. J. (1973). Methods of handling ties in the Wilcoxon signed-rank test. *Journal of the American Statistical Association* **68**, 985–988.

Conover, W. J. (1974). Some reasons for not using the Yates continuity correction on 2 × 2 contingency tables. *Journal of the American Statistical Association* **69**, 374–376.

Cook, R. D. (1977). Detection of influential observations in linear regression. *Technometrics* **19**, 15–18.

Cook, R. D. (1979). Influential observations in linear regression. *Journal of the American Statistical Association* **74**, 169–174.

Cook, R. D. and Weisberg, S. (1983). Diagnostics for heteroscedasticity in regression. *Biometrika* **70**, 1–10.

Cornfield, J. and Tukey, J. W. (1956). Average values of mean squares in factorials. *Annals of Mathematical Statistics* **27**, 907–949.

Creasy, M. A. (1956). Confidence limits for the gradient in the linear functional relationship. *Journal of the Royal Statistical Society, Series B* **18**, 65–69.

Cressie, N. (1981). Transformations and the jackknife. *Journal of the Royal Statistical Society, Series B* **43**, 177–182.

Cureton, E. E. (1967). The normal approximation to the signed-rank sampling distribution when zero differences are present. *Journal of the American Statistical Association* **62**, 1068–1069.

David, H. A. (1952). Upper 5 and 1% points of the maximum F-ratio. *Biometrika* **39**, 422–424.

Davies, R. B. and Hutton, B. (1975). The effect of errors in the independent variables in linear regression. *Biometrika* **62**, 383–391.

DeGroot, M. (1973). Doing what comes naturally: interpreting a tail area as a posterior probability or as a likelihood ratio. *Journal of the American Statistical Association* **68**, 966–969.

Deming, W. E. (1943). *Statistical Adjustment of Data*. Wiley, New York.

Devlin, S. J., Gnanadesikan, R., and Kettenring, J. R. (1975). Robust estimation and outlier detection with correlation coefficients. *Biometrika* **62**, 531–545.

Dixon, D. O. and Duncan, D. B. (1975). Minimum Bayes risk t-intervals for multiple comparisons. *Journal of the American Statistical Association* **70**, 822–831.

Dixon, W. J. (1944). Further contributions to the problem of serial correlation. *Annals of Mathematical Statistics* **15**, 119–144.

Doksum, K. A. and Wong, C.-W. (1983). Statistical tests based on transformed data. *Journal of the American Statistical Association* **78**, 411–417.

Draper, N. and Smith, H. (1981). *Applied Regression Analysis*, Second Edition. Wiley, New York.

Duan, N. (1981). Consistency of residual distribution function. Working Draft No. 801-1-HHS (016B-8001), Rand Corporation, Santa Monica, California.

Duncan, G. T. and Layard, M. W. J. (1973). A Monte-Carlo study of asymptotically robust tests for correlation coefficients. *Biometrika* **60**, 551–558.

Dunn, O. J. (1961). Multiple comparisons among means. *Journal of the American Statistical Association* **56**, 52–64.

Dunn, O. J. (1964). Multiple comparisons using rank sums. *Technometrics* **6**, 241–252.

Dunnett, C. W. (1980a). Pairwise multiple comparisons in the homogeneous variance, unequal sample size case. *Journal of the American Statistical Association* **75**, 789–795.

Dunnett, C. W. (1980b). Pairwise multiple comparisons in the unequal variance case. *Journal of the American Statistical Association* **75**, 796–800.

Durbin, J. (1959). A note on the application of Quenouille's method of bias reduction to the estimation of ratios. *Biometrika* **46**, 477–480.

Durbin, J. and Watson, G. S. (1950). Testing for serial correlation in least squares regression. I. *Biometrika* **37**, 409–428.

Durbin, J. and Watson, G. S. (1951). Testing for serial correlation in least squares regression. II. *Biometrika* **38**, 159–178.

Durbin, J. and Watson, G. S. (1971). Testing for serial correlation in least squares regression. III. *Biometrika* **58**, 1–19.

Dwass, M. (1960). Some k-sample rank-order tests. *Contributions to Probability and Statistics*, (I. Olkin, S. G. Ghurye, W. Hoeffding, W. G. Madow, and H. B. Mann, Eds.). Stanford University Press, Stanford, pp. 198–202.

Efron, B. (1969). Student's t-test under symmetry conditions. *Journal of the American Statistical Association* **64**, 1278–1302.

Efron, B. (1979). Bootstrap methods: another look at the jackknife. *Annals of Statistics* **7**, 1–26.

Efron, B. (1981). Nonparmetric estimates of standard error: the jackknife, the bootstrap and other methods. *Biometrika* **68**, 589–599.

Efron, B. (1982). *The Jackknife, the Bootstrap, and Other Resampling Plans.* Society for Industrial and Applied Mathematics, Philadelphia.

Efron, B. and Morris, C. (1971). Limiting the risk of Bayes and empirical Bayes estimators – Part I: the Bayes case. *Journal of the American Statistical Association* **66**, 807–815.

Efron, B. and Morris, C. (1972). Limiting the risk of Bayes and empirical Bayes estimators – Part II: the empirical Bayes case. *Journal of the American Statistical Association* **67**, 130–139.

Efron, B. and Morris, C. (1973a). Stein's estimation rule and its competitors – an empirical Bayes approach. *Journal of the American Statistical Association* **68**, 117–130.

Efron, B. and Morris, C. (1973b). Combining possibly related estimation problems. *Journal of the Royal Statistical Society, Series B* **35**, 379–402.

Efron, B. and Morris, C. (1975). Data analysis using Stein's estimator and its generalizations. *Journal of the American Statistical Association* **70**, 311–319.

Eisenhart, C. (1947). The assumptions underlying the analysis of variance. *Biometrics* **3**, 1–21.

Eisenhart, C. and Solomon, H. (1947). Significance of the largest of a set of sample estimates of variance. *Techniques of Statistical Analysis*, (C. Eisenhart, M. W. Hastay, and W. A. Wallis, Eds.). McGraw-Hill, New York, pp. 383–394.

Faith, R. E. (1976). Minimax Bayes set and point estimators of a multivariate normal mean. Technical Report No. 66, Department of Statistics, University of Michigan.

Fieller, E. C. (1932). The distribution of the index in a normal bivariate population. *Biometrika* **24**, 428–440.

Fieller, E. C. (1940). The biological standardization of insulin. *Royal Statistical Society, Supplement* (superseded by *Series B*) **7**, 1–64.

Fieller, E. C. (1954). Some problems in interval estimation. *Journal of the Royal Statistical Society, Series B* **16**, 175–185.

Finch, D. J. (1950). The effect of non-normality on the z-test, when used to compare the variances in two populations. *Biometrika* **37**, 186–189.

Finney, D. J. (1978). *Statistical Method in Biological Assay*. Griffin, London; Macmillan, New York.

Finney, D. J., Latscha, R., Bennett, B. M., and Hsu, P. (1963). *Tables for Testing Significance in a* 2×2 *Contingency Table*. University Press, Cambridge.

Fisher, R. A. (1921). On the "probable error" of a coefficient of correlation deduced from a small sample. *Metron* **1** (4), 3–32.

Fisher, R. A. (1934). *Statistical Methods for Research Workers*, Fifth Edition. Oliver & Boyd, Edinburgh.

Fisher, R. A. (1935). *The Design of Experiments*. Oliver & Boyd, Edinburgh.

Fraser, D. A. S. (1957). *Nonparametric Methods in Statistics*. Wiley, New York.

Freund, J. E. and Ansari, A. R. (1957). Two-way rank-sum tests for variances. Technical Report No. 34, Department of Statistics and Statistical Laboratory, Virginia Polytechnic Institute.

Friedman, M. (1937). The use of ranks to avoid the assumption of normality implicit in the analysis of variance. *Journal of the American Statistical Association* **32**, 675–701.

Fuller, W. A. and Rao, J. N. K. (1978). Estimation for a linear regression model with unknown diagonal covariance matrix. *Annals of Statistics* **6**, 1149–1158.

Gabriel, K. R. (1978). A simple method of multiple comparisons of means. *Journal of the American Statistical Association* **73**, 724–729.

Gastwirth, J. L. and Rubin, H. (1971). Effect of dependence on the level of some one-sample tests. *Journal of the American Statistical Association* **66**, 816–820.

Gastwirth, J. L. and Rubin, H. (1975). The behavior of robust estimators on dependent data. *Annals of Statistics* **3**, 1070–1100.

Gautschi, W. (1959). Some remarks on Herbach's paper, "Optimum nature of the F-test for Model II in the balanced case." *Annals of Mathematical Statistics* **30**, 960–963.

Gayen, A. K. (1949). The distribution of "Student's" t in random samples of any size drawn from non-normal universes. *Biometrika* **36**, 353–369.

Gayen, A. K. (1950a). The distribution of the variance ratio in random samples of any size drawn from non-normal universes.

Biometrika **37**, 236–255.

Gayen, A. K. (1950b). Significance of difference between the means of two non-normal samples. *Biometrika* **37**, 399–408.

Gayen, A. K. (1951). The frequency distribution of the product-moment correlation coefficient in random samples of any size drawn from non-normal universes. *Biometrika* **38**, 219–247.

Geary, R. C. (1930). The frequency distribution of the quotient of two normal variates. *Journal of the Royal Satistical Society* (superseded by *Series A*) **93**, 442–446.

Geary, R. C. (1936). The distribution of "Student's" ratio for non-normal samples. *Journal of the Royal Statistical Society Supplement* (superseded by *Series B*) **3**, 178–184.

Geary, R. C. (1947). Testing for normality. *Biometrika* **34**, 209–242.

Geisser, S. and Greenhouse, S. W. (1958). An extension of Box's results on the use of the F distribution in multivariate analysis. *Annals of Mathematical Statistics* **29**, 885–891.

Gibbons, J. D. (1971). *Nonparametric Statistical Inference.* McGraw-Hill, New York.

Gibbons, J. D. and Pratt, J. W. (1975). P-values: interpretations and methodology. *American Statistician* **29**, 20–25.

Gleser, L. J. (1981). Estimation in a multivariate "errors in variables" regression model: large sample results. *Annals of Statistics* **9**, 24–44.

Graybill, F. A. (1961). *An Introduction to Linear Statistical Models.* McGraw-Hill, New York.

Graybill, F. A. (1976). *Theory and Application of the Linear Model.* Duxbury Press, North Scituate, Mass.

Graybill, F. A. and Bowden, D. C. (1967). Linear segment confidence bands for simple linear models. *Journal of the American Statistical Association* **62**, 403–408.

Graybill, F. A. and Deal, R. B. (1959). Combining unbiased estimtors. *Biometrics* **15**, 543–550.

Grizzle, J. E. (1967). Continuity correction in the χ^2-test for 2 × 2 tables. *American Statistician* **21** (4), 28–32.

Grizzle, J. E. (1969). Letter to the Editor. *American Statistician* **23** (2), 35.

Gross, A. M. (1976). Confidence interval robustness with long-tailed symmetric distributions. *Journal of the American Statistical Association* **71**, 409–416.

Gross, A. M. (1977). Confidence intervals for bisqure regression estimates. *Journal of the American Statistical Association* **72**, 341–354.

Hahn, G. J. and Hendrickson, R. W. (1971). A table of percentage points of the distribution of the largest absolute value of k Student t variates and its applications. *Biometrika* **58**, 323–332.

Hajek, J. and Šidák, Z. (1967). *Theory of Rank Tests.* Academic Press, New York.

Hampel, F. R. (1974). The influence curve and its role in robust estimation. *Journal of the American Statistical Association* **69**, 383–393.

Hannan, E. J. (1960). *Time Series Analysis.* Methuen, London.

Harter, H. L. (1960). Tables of range and studentized range. *Annals of Mathematical Statistics* **31**, 1122–1147.

Harter, H. L. (1961). Expected values of normal order statistics. *Biometrika* **48**, 151–165.

Harter, H. L. (1969a). *Order Statistics and Their Use in Testing and Estimation*, Vol. 1. *Tests Based on Range and Studentized Range of Samples from a Normal Population.* Aerospace Research Laboratories. (Available from U.S. Government Printing Office, Washington, DC 20402.)

Harter, H. L. (1969b). *Order Statistics and Their Use in Testing and Estimation*, Vol. 2. *Estimates Based on Order Statistics of Samples from Various Populations.* Aerospace Research Laboratories. (Available from U.S. Government Printing Office, Washington, DC 20402.)

Hartley, H. O. (1950). The maximum F-ratio as a short-cut test for heterogeneity of variance. *Biometrika* **37**, 308–312.

Hartley, H. O. and Ross, A. (1954). Unbiased ratio estimators. *Nature* (London) **174**, 270–271.

Harvard Computation Laboratory (1955). *Tables of Cumulative Binomial Probability Distribution.* Harvard University Press, Cambridge.

Hayter, A. J. (1984). A proof of the conjecture that the Tukey-Kramer multiple comparisons procedure is conservative. *Annals of Statistics* **12**, 61–75.

Herbach, L. H. (1959). Properties of Model II-type analysis of variance tests, A: optimum nature of the F-test for Model II in the balanced case. *Annals of Mathematical Statistics* **30**, 939–959.

Hinkley, D. V. (1969). On the ratio of two correlated normal random variables. *Biometrika* **56**, 635–639.

Hinkley, D. V. (1975). On power transformations to symmetry. *Biometrika* **62**, 101–111.

Hinkley, D. V. (1978). Improving the jackknife with special reference to correlation estimation. *Biometrika* **65**, 13–21.

Hinkley, D. V. and Wang, H.-L. (1980). A trimmed jackknife. *Journal of the Royal Statistical Society, Series B* **42**, 347–356.

Hoadley, B. (1970). A Bayesian look at inverse regression. *Journal of the American Statistical Association* **65**, 356–369.

Hoaglin, D. C. and Welsch, R. E. (1978). The hat matrix in regression and ANOVA. *American Statistician* **32**, 17–22. Corrigenda: **32** (1978), 146.

Hochberg, Y. (1974). Some conservative generalizations of the T-method in simultaneous inference. *Journal of Multivariate Analysis* **4**, 224–234.

Hodges, J. L., Jr. and Lehmann, E. L. (1951). Some applications of the Cramér-Rao inequality. *Proceedings of the Second Berkeley Symposium on Mathematical Statistics and Probability*, (J. Neyman, Ed.). University of California Press, Berkeley, pp. 13–22.

Hodges, J. L., Jr. and Lehmann, E. L. (1961). Comparison of the normal scores and Wilcoxon tests. *Proceedings of the Fourth Berkeley Symposium on Mathematical Statistics and Probability*, Vol. I, (J. Neyman, Ed.). University of California Press, Berkeley, pp. 307–317.

Hodges, J. L., Jr. and Lehmann, E. L. (1962). Rank methods for combination of independent experiments in analysis of variance. *Annals of Mathematical Statistics* **33**, 482–497.

Hodges, J. L., Jr. and Lehmann, E. L. (1963). Estimates of location based on rank tests. *Annals of Mathematical Statistics* **34**, 598–611.

Hoeffding, W. (1948). A class of statistics with asymptotically normal distibution. *Annals of Mathematical Statistics* **19**, 293–325.

Hollander, M., Pledger, G., and Lin, P.-E. (1974). Robustness of the Wilcoxon test to a certain dependency between samples. *Annals of Statistics* **2**, 177–181.

Hollander, M. and Wolfe, D. A. (1973). *Nonparametric Statistical Methods.* Wiley, New York.

Hotelling, H. (1953). New light on the correlation coefficient and its transforms. *Journal of the Royal Statistical Society, Series B* **15**, 193–225.

Hsu, P. L. (1938). Contributions to the theory of "Student's" t-test as applied to the problem of two samples. *Statistical Research Memoirs*, Vol. II, (J. Neyman and E. S. Pearson, Eds.). Department of Statistics, Univeristy College, London, pp. 1–24.

Huber, P. J. (1973). Robust regression: asymptotics, conjectures and Monte Carlo. *Annals of Statistics* **1**, 799–821.

Huber, P. J. (1975). Robustness and designs. *A Survey of Statistical Designs and Linear Models*, (J. N. Srivastava, Ed.). North-Holland, Amsterdam; American Elsevier, New York.

Huber, P. J. (1977). *Robust Statistical Procedures*. Society for Industrial and Applied Mathematics, Philadelphia.

Huber, P. J. (1981). *Robust Statistics*. Wiley, New York.

Hunter, W. G. and Lamboy, W. F. (1981). A Bayesian analysis of linear calibration. *Technometrics* **23**, 323–328.

Hutchinson, M. C. (1971). A Monte Carlo comparison of some ratio estimators. *Biometrika* **58**, 313–321.

Huynh, H. and Feldt, L. S. (1970). Conditions under which mean square ratios in repeated measurements designs have exact *F*-distributions. *Journal of the American Statistical Association* **65**, 1582–1589.

Huynh, H. and Feldt, L. S. (1980). Performance of traditional *F* tests in repeated measures designs under covariance heterogeneity. *Communications in Statistics, Part A* **9**, 61–74.

Iman, R. L., Quade, D., and Alexander, D. A. (1975). Exact probability levels for the Kruskal-Wallis tests. *Selected Tables in Mathematical Statistics*, Vol. 3, (H. L. Harter and D. B. Owen, Eds.). American Mathematical Society, Providence, pp. 329–384.

Irwin, J. O. (1935). Tests of significance for differences between percentages based on small numbers. *Metron* **12**, 83–94.

Jacquez, J. A., Mather, F. J., and Crawford, C. A. (1968). Linear regression with nonconstant, unknown error variances: sampling experiments with least squares, weighted least squares and maximum likelihood estimators. *Biometrics* **24**, 607–626.

James, W. and Stein, C. (1961). Estimation with quadratic loss. *Proceedings of the Fourth Berkeley Symposium on Mathematical*

Statistics and Probability, Vol. I, (J. Neyman, Ed.). Univeristy of California Press, Berkeley, pp. 361–379.

Jonckheere, A. R. (1954). A distribution-free k-sample test against ordered alternatives. *Biometrika* **41**, 133–145.

Joshi, V. M. (1967). Inadmissibility of the usual confidence sets for the mean of a multivariate normal population. *Annals of Mathematical Statistics* **38**, 1868–1875.

Jurečková, J. (1977). Asymptotic relations of M-estimates and R-estimates in linear regression. *Annals of Statistics* **5**, 464–472.

Kelly, G. (1984). The influence function in the errors in variables problem. *Annals of Statistics* **12**, 87–100.

Kempthorne, O. (1952). *The Design and Analysis of Experiments.* Wiley, New York.

Kendall, M. G. and Stuart, A. (1961). *The Advanced Theory of Statistics*, Vol. 2. Griffin, London; Hafner, New York.

Klotz, J. H. (1962). Nonparametric tests for scale. *Annals of Mathematical Statistics* **33**, 498–512.

Klotz, J. H. (1966). The Wilcoxon, ties, and the computer. *Journal of the American Statistical Association* **61**, 772–787.

Klotz, J. H., Milton, R. C. and Zacks, S. (1969). Mean square efficiency of estimators of variance components. *Journal of the American Statistical Association* **64**, 1383–1402.

Koenker, R. and Bassett, G., Jr. (1978). Regression quantiles. *Econometrica* **46**, 33–50.

Koziol, J. A. and Reid, N. (1977). On the asymptotic equivalence of two ranking methods for K-sample linear rank statistics. *Annals of Statistics* **5**, 1099–1106.

Kraft, C. H. and van Eeden, C. (1968). *A Nonparametric Introduction to Statistics.* Macmillan, New York.

Kramer, C. Y. (1956). Extension of multiple range tests to group means with unequal numbers of replications. *Biometrics* **12**, 307–310.

Kruskal, W. H. (1968). When are Gauss-Markov and least squares estimators identical? A coordinate-free approach. *Annals of Mathematical Statistics* **39**, 70–75.

Kruskal, W. H. and Wallis, W. A. (1952). Use of ranks in one-criterion variance analysis. *Journal of the American Statistical Association* **47**, 583–621. Errata: **48** (1953), 907–911.

Krutchkoff, R. G. (1967). Classical and inverse regression methods of calibration. *Technometrics* **9**, 425–439.

Kurtz, T. E. (1956). An extension of a multiple comparison procedure. Ph.D. dissertation, Princeton University.

LaMotte, L. R. (1973). On non-negative quadratic unbiased estimation of variance components. *Journal of the American Statistical Association* **68**, 728–730.

Layard, M. W. J. (1973). Robust large-sample tests for homogeneity of variances. *Journal of the American Statistical Association* **68**, 195–198.

Lehmann, E. L. (1975). *Nonparametrics: Statistical Methods Based on Ranks*. Holden-Day, San Francisco.

Levene, H. (1960). Robust tests for equality of variances. *Contributions to Probability and Statistics*, (I. Olkin, S. G. Ghurye, W. Hoeffding, W. G. Madow, and H. B. Mann, Eds.). Stanford University Press, Stanford, pp. 278–292.

Lieberman, G. J., Miller, R. G., Jr., and Hamilton, M. A. (1967). Unlimited simultaneous discrimination intervals in regression. *Biometrika* **54**, 133–145. Correction: **58** (1971), 687.

Lindley, D. V. (1947). Regression lines and the linear functional relationship. *Journal of the Royal Statistical Society, Supplement* (superseded by *Series B*) **9**, 218–244.

Lindley, D. V. (1962). Discussion on Professor Stein's paper. *Journal of the Royal Statistical Society, Series B* **24**, 285–287.

Lindley, D. V. and El-Sayyad, G. M. (1968). The Bayesian estimation of a linear functional relationship. *Journal of the Royal Statistical Society, Series B* **30**, 190–202.

Madansky, A. (1959). The fitting of straight lines when both variables are subject to error. *Journal of the American Statistical Association* **54**, 173–205.

Malley, J. D. (1982). Simultaneous confidence intervals for ratios of normal means. *Journal of the American Statistical Association* **77**, 170–176.

Mann, H. B. and Whitney,D. R. (1947). On a test of whether one of two random variables is stochastically larger than the other. *Annals of Mathematical Statistics* **18**, 50–60.

Mantel, N. and Greenhouse, S. W. (1968). What is the continuity correction? *American Statistician* **22** (5), 27–30.

Marsaglia, G. (1965). Ratios of normal variables and ratios of sums of uniform variables. *Journal of the American Statistical Association* **60**, 193–204.

Mehra, K. L. and Sarangi, J. (1967). Asymptotic efficiency of certain rank tests for comparative experiments, *Annals of Mathematical Statistics* **38**, 90–107.

Mickey, M. R. (1959). Some finite population unbiased ratio and regression estimators. *Journal of the American Statistical Association* **54**, 594–612.

Miller, R. G., Jr. (1964). A trustworthy jackknife. *Annals of Mathematical Statistics* **35**, 1594–1605.

Miller, R. G., Jr. (1968). Jackknifing variances. *Annals of Mathematical Statistics* **39**, 567–582.

Miller, R. G., Jr. (1974a). The jackknife – a review. *Biometrika* **61**, 1–15.

Miller, R. G., Jr. (1974b). An unbalanced jackknife. *Annals of Statistics* **2**, 880–891.

Miller, R. G., Jr. (1981). *Simultaneous Statistical Inference*, Second Edition. Springer-Verlag, New York.

Miller, R. G., Jr. (1985). Multiple comparisons. *Encyclopedia of Statistical Sciences*, Vol. 5, (S. Kotz and N. L. Johnson, Eds.). Wiley, New York.

Milliken, G. A. and Graybill, F. A. (1970). Extensions of the general linear hypothesis model. *Journal of the American Statistical Association* **65**, 797–807.

Mood, A. M. (1950). *Introduction to the Theory of Statistics.* McGraw-Hill, New York.

Morris, C. N. (1983). Parametric empirical Bayes inference: theory and applications. *Journal of the American Statistical Association* **78**, 47–55.

Moses, L. E. (1963). Rank tests of dispersion. *Annals of Mathematical Statistics* **34**, 973–983.

Moses, L. E. (1978). Charts for finding upper percentage points of Student's t in the range.01 to .00001. *Communications in Statistics, Part B* **7**, 479–490.

Mosteller, F. and Tukey, J. W. (1968). Data analysis, including statistics. *Handbook of Social Psychology*, (G. Lindzey and E. Aronson, Eds.). Addison-Wesley, Reading, Mass.

Mosteller, F. and Tukey, J. W. (1977). *Data Analysis and Regression.* Addison-Wesley, Reading, Mass.

Murdock, G. R. and Williford, W. O. (1977). Tables for obtaining optimal confidence intervals involving the chi-square distribution. *Selected Tables in Mathematical Statistics*, Vol. 5, (D. B. Owen and R. E. Odeh, Eds.). American Mathematical Society, Providence.

Nair, K. R. and Banerjee, K. S. (1942/44). A note on fitting of straight lines if both variables are subject to error. *Sankhyā* **6**, 331.

Nair, K. R. and Shrivastava, M. P. (1942/44). On a simple method of curve fitting. *Sankhyā* **6**, 121–132.

Nelson, L. S. (1977). Tables for testing ordered alternatives in an analysis of variance. *Biometrika* **64**, 335–338.

Nemenyi, P. (1963). Distribution-free multiple comparisons. Ph.D. dissertation, Princeton University.

Nieto do Pascual, J. (1961). Unbiased ratio estimators in stratified sampling. *Journal of the American Statistical Association* **56**, 70–87.

Noether, G. E. (1967). *Elements of Nonparametric Statistics.* Wiley, New York.

Norwood, T. E. and Hinkelmann, K. (1977). Estimating the common mean of several normal populations. *Annals of Statistics* **5**, 1047–1050.

Olkin, I. and Vaeth, M. (1981). Maximum likelihood estimation in a two-way analysis of variance with correlated errors in one classification. *Biometrika* **68**, 653–660.

Owen, D. B. (1962). *Handbook of Statistical Tables.* Addison-Wesley, Reading, Mass.

Page, E. B. (1963). Ordered hypotheses for multiple treatments: a significance test for linear ranks. *Journal of the American Statistical Association* **58**, 216–230.

Pascual, J. N. (1961). See Nieto de Pascual, J. (1961).

Paulson, E. (1942). A note on the estimation of some mean values for a bivariate distribution. *Annals of Mathematical Statistics* **13**, 440–445.

Pearson, E. S. (1929). The distribution of frequency constants in small samples from non-normal symmetrical and skew populations. *Biometrika* **21**, 259–286.

Pearson, E. S. (1931). The analysis of variance in cases of non-normal variation. *Biometrika* **23**, 114–133.

Pearson, E. S. and Hartley, H. O. (1970). *Biometrika Tables for Statisticians*, Vol. I, Third Edition. University Press, Cambridge.

Pearson, E. S. and Hartley, H. O. (1972). *Biometrika Tables for Statisticians*, Vol. II. University Press, Cambridge.

Pearson, E. S. and Please, N. W. (1975). Relation between the shape of population distribution and the robustness of four simple test statistics. *Biometrika* **62**, 223–241.

Pierce, D. A. and Gray, R. J. (1982). Testing normality of errors in regression models. *Biometrika* **69**, 233–236.

Pierce, D. A. and Kopecky, K. J. (1979). Testing goodness of fit for the distribution of errors in regression models *Biometrika* **66**, 1–5.

Pitman, E. J. G. (1937). Significance tests which may be applied to samples from any populations. *J. Roy. Statist. Soc.* (superseded by *Series B*) **4**, 119-130.

Pitman, E. J. G. (1938). Significance tests which may be applied to samples from any populations. III. The analysis of variance test. *Biometrika* **29**, 322–335.

Portnoy, S. (1971). Formal Bayes estimation with application to a random effects model. *Annals of Mathematical Statistics* **42**, 1379–1402.

Potthoff, R. F. (1963). Use of the Wilcoxon statistic for a generalized Behrens-Fisher problem. *Annals of Mathematical Statistics* **34**, 1596–1599.

Pratt, J. W. (1959). Remarks on zeroes and ties in the Wilcoxon signed ranks procedure. *Journal of the American Statistical Association* **54**, 655–667.

Pratt, J. W. and Gibbons, J. D. (1981). *Concepts of Nonparametric Theory*. Springer-Verlag, New York.

Pukelsheim, F. (1981). On the existence of unbiased nonnegative estimates of variance covariance components. *Annals of Statistics* **9**, 293–299.

Putter, J. (1955). The treatment of ties in some nonparametric tests. *Annals of Mathematical Statistics* **26**, 368–386.

Quenouille, M. H. (1949). Approximate tests of correlation in time-series. *Journal of the Royal Statistical Society, Series B* **11**, 68–84.

Quenouille, M. H. (1956). Notes on bias in estimation. *Biometrika* **43**, 353–360.

Rahe, A. J. (1974). Tables of critical values for the Pratt matched pair signed rank statistic. *Journal of the American Statistical Association* **69**, 368–373.

Rankin, N. O. (1974). The harmonic mean method for one-way and two-way analyses of variance. *Biometrika* **61**, 117–122.

Rao, C. R. (1970). Estimation of heteroscedastic variances in linear models. *Journal of the American Statistical Association* 65, 161–172.

Rao,C. R. (1971). Estimation of variance and covariance components – MINQUE theory. *Journal of Multivariate Analysis* **1**, 257–275.

Rao, C. R. (1972). Estimating variance and covariance components in linear models. *Journal of the American Statistical Association* **67**, 112–115.

Rao, C. R. (1973). *Linear Statistical Inference and Its Applications*, Second Edition. Wiley, New York.

Rao, J. N. K. (1965). A note on estimation of ratios by Quenouille's method. *Biometrika* **52**, 647–649.

Rao, J. N. K. (1969). Ratio and regression estimators. *New Developments in Survey Sampling*, (N. L. Johnson and H. Smith, Jr., Eds.). Wiley-Interscience, New York, pp. 213–234.

Rao, J. N. K. and Beegle, L. D. (1967). A Monte Carlo study of some ratio estimators. *Sankhyā B* **29**, 47–56.

Rao, J. N. K. and Webster, J. (1966). On two methods of bias reduction in the estimation of ratios. *Biometrika* **53**, 571–577.

Rao, P. S. R. S. (1969). Comparison of four ratio-type estimates under a model. *Journal of the American Statistical Association* **64**, 574–580.

Rao, P. S. R. S., Kaplan, J., and Cochran, W. G. (1981). Estimators for the one-way random effects model with unequal error variances. *Journal of the American Statistical Association* **76**, 89–97.

Rao, P. S. R. S. and Rao, J. N. K. (1971). Small sample results for ratio estimators. *Biometrika* **58**, 625–630.

Ringland, J. T. (1983). Robust multiple comparisons. *Journal of the American Statistical Association* **78**, 145–151.

Rosenblatt, J. R. and Spiegelman, C. H. (1981). Discussion. *Technometrics* **23**, 329–333.

Ruppert, D. and Carroll, R. J. (1980). Trimmed least squares estimation in the linear model. *Journal of the American Statistical Association* **75**, 828–838.

Satterthwaite, F. E. (1946). An approximate distribution of estimates of variance components. *Biometrics Bulletin* (superseded by *Biometrics*) **2**, 110–114.

Scheffé, H. (1943). On solutions of the Behrens-Fisher problem, based on the *t*-distribution. *Annals of Mathematical Statistics* **14**, 35–44.

Scheffé, H. (1944). A note on the Behrens-Fisher problem. *Annals of Mathematical Statistics* **15**, 430–432.

Scheffé, H. (1953). A method for judging all contrasts in the analysis of variance. *Biometrika* **40**, 87–104.

Scheffé, H. (1959). *The Analysis of Variance*. Wiley, New York.

Scheffé, H. (1970a). Multiple testing versus multiple estimation. Improper confidence sets. Estimation of directions and ratios. *Annals of Mathematical Statistics* **41**, 1–29.

Scheffé, H. (1970b). Practical solutions of the Behrens-Fisher problem. *Journal of the American Statistical Association* **65**, 1501–1508.

Scheffé, H. (1973). A statistical theory of calibration. *Annals of Statistics* **1**, 1–37.

Schrader, R. M. and Hettmansperger, T. P. (1980). Robust analysis of variance based upon a likelihood ratio criterion. *Biometrika* **67**, 93–101.

Scott, A. and Wu, C.-F. (1981). On the asymptotic distribution of ratio and regression estimators. *Journal of the American Statistical Association* **76**, 98–102.

Searle, S. R. (1971). *Linear Models*. Wiley, New York.

Sen, P. K. (1968). Estimates of the regression coefficient based on Kendall's tau. *Journal of the American Statistical Association* **63**, 1379–1389.

Serfling, R. J. (1968). The Wilcoxon two-sample statistic on strongly mixing processes. *Annals of Mathematical Statistics* **39**, 1202–1209.

Shapiro, S. S. and Francia, R. S. (1972). An approximate analysis of variance test for normality. *Journal of the American Statistical Association* **67**, 215–216.

Shapiro, S. S. and Wilk, M. B. (1965). An analysis of variance test for normality (complete samples). *Biometrika* **52**, 591–611.

Shapiro, S. S., Wilk, M. B., and Chen, H. J. (1968). A comparative study of various tests for normality. *Journal of the American Statistical Association* **63**, 1343–1372.

Shorack, G. R. (1966). Graphical procedures for using distribution free methods in the estimation of relative potency in dilution

(-direct) assays. *Biometrics* **22**, 610–619.

Shorack, G. R. (1969). Testing and estimating ratios of scale parameters. *Journal of the American Statistical Association* **64**, 999–1013.

Šidák, Z. (1967). Rectangular confidence regions for the means of multivariate normal distributions. *Journal of the American Statistical Association* **62**, 626–633.

Siegel, A. F. (1982). Robust regression using repeated medians. *Biometrika* **69**, 242–244.

Siegel, S. and Tukey, J. W. (1960). A nonparametric sum of ranks procedure for relative spread in unpaired samples. *Journal of the American Statistical Association* **55**, 429–445. Correction: **56** (1961), 1005.

Skillings, J. H. and Mack, G. A. (1981). On the use of a Friedman-type statistic in balanced and unbalanced block designs. *Technometrics* **23**, 171–177.

Spjøtvoll, E. (1967). Optimum invariant tests in unbalanced variance components models. *Annals of Mathematical Statistics* **38**, 422–428.

Spjøtvoll, E. and Stoline, M. R. (1973). An extension of the T-method of multiple comparisons to include the cases with unequal sample sizes. *Journal of the American Statistical Association* **68**, 975–978.

Sprent, P. (1966). A generalized least-squares approch to linear functional relationships. *Journal of the Royal Statistical Society, Series B* **28**, 278–288.

Steel, R. G. D. (1960). A rank sum test for comparing all pairs of treatments. *Technometrics* **2**, 197–207.

Stein, C. M. (1956). Inadmissibility of the usual estimator for the mean of a multivariate normal distribution. *Proceedings of the*

Third Berkeley Symposium on Mathematical Statistics and Probability, Vol. I, (J. Neyman, Ed.). University of California Press, Berkeley, pp. 197–206.

Stein, C. M. (1962). Confidence sets for the mean of a multivariate normal distribution. *Journal of the Royal Statistical Society, Series B* **24**, 265–285.

Stein, C. M. (1964). Inadmissibility of the usual estimator for the variance of a normal distribution with unknown mean. *Annals of the Institute of Statistical Mathematics* **16**, 155–160.

Stigler, S. M. (1977). Do robust estimators work with real data? *Annals of Statistics* **5**, 1055–1078.

Student (1908). The probable error of a mean. *Biometrika* **6**, 1–25.

Szatrowski, T. H. and Miller, J. J. (1980). Explicit maximum likelihood estimates from balanced data in the mixed model of the analysis of variance. *Annals of Statistics* **8**, 811–819.

Tamhane, A. C. (1979). A comparison of procedures for multiple comparisons of means with unequal variances. *Journal of the American Statistical Association* **74**, 471–480.

Terpstra, T. J. (1952). The asymptotic normality and consistency of Kendall's test against trend, when ties are present in one ranking. *Indagationes Mathematicae* **14**, 327–333.

Theil, H. (1950). A rank-invariant method of linear and polynomial regression. I. *Indagationes Mathematicae* **12**, 85–91.

Tin, M. (1965). Comparison of some ratio estimators. *Journal of the American Statistical Association* **60**, 294–307.

Tukey, J. W. (1949). One degree of freedom for non-additivity. *Biometrics* **5**, 232–242.

Tukey, J. W. (1953). The problem of multiple comparisons. Unpublished manuscript.

Tukey, J. W. (1955). Query 113: Answer. *Biometrics* **11**, 111–113.

Tukey, J. W. (1957). On the comparative anatomy of transformations. *Annals of Mathematical Statistics* **28**, 602–632.

Tukey, J. W. (1958). Bias and confidence in not-quite large samples (abstract). *Annals of Mathematical Statistics* **29**, 614.

Tukey, J. W. (1977). *Exploratory Data Analysis*. Addison-Wesley, Reading, Mass.

Tukey, J. W. and McLaughlin, D. H. (1963). Less vulnerable confidence and significance procedures for location based on a single sample: trimming/winsorization 1. *Sankhyā, Series A* **25**, 331–352.

Wald, A. (1940). Fitting of straight lines if both variables are subject to error. *Annals of Mathematical Statistics* **11**, 284–300.

Walker, H. M. and Lev, J. (1953). *Statistical Inference*. Holt, New York.

Walters, D. E. and Rowell, J. G. (1982). Comments on a paper by I. Olkin and M. Vaeth on two-way analysis of variance with correlated errors. *Biomerika* **69**, 664–666.

Watson, G. S. (1967). Linear least squares regression. *Annals of Mathematical Statistics* **38**, 1679–1699.

Welch,B. L. (1937). On the z-test in randomized blocks and Latin squares. *Biometrika* **29**, 21–52.

Welch, B. L. (1947). The generalization of "Student's" problem when several different population variances are involved. *Biometrika* **34**, 28–35.

Welch, B. L. (1949). Further note on Mrs. Aspin's tables and on certain approximations to the tabled function. *Biometrika* **36**, 293–296.

Wilcoxon, F. (1945). Individual comparisons by ranking methods. *Biometrics Bulletin* (superseded by *Biometrics*) **1**, 80–83.

Wilk, M. B. and Gnanadesikan, R. (1968). Probability plotting methods for the analysis of data. *Biometrika* **55**, 1–17.

Wolter, K. M. and Fuller, W. A. (1982). Estimation of the quadratic errors-in-variables model. *Biometrika* **69**, 175–182.

Working, H. and Hotelling, H. (1929). Application of the theory of error to the interpretation of trends. *Journal of the American Statistical Association, Supplement (Proceedings)* **24**, 73–85.

Yates, F. (1934). Contingency tables involving small numbers and the χ^2 test. *Journal of the Royal Statistical Society, Supplement* (superseded by *Series B*) **1**, 217–235.

Yuen, K. K. (1974). The two-sample trimmed t for unequal population variances. *Biometrika* **61**, 165–170.

Yuen, K. K. and Dixon, W. J. (1973). The approximate behaviour and performance of the two-sample trimmed t. *Biometrika* **60**, 369–374.

Zyskind, G. (1967). On canonical forms, non-negative covariance matrices and best and simple least squares linear estimators in linear models. *Annals of Mathematical Statistics* **38**, 1092–1109.

AUTHOR INDEX

Abelson, 78–80, 81, 88, 91, 131.
Ablashi, 66.
Abraham, 127.
Alderman, 65.
Alexander, 85.
Amemiya, 211, 212.
Andersen, 80, 81, 89, 135, 142, 269, 271, 272, 274, 276, 277.
Anderson, R.L., 35, 144.
Anderson, T., 161.
Anderson, T.W., 35, 175, 228.
Andrews, 28, 32, 198, 201, 202, 203.
Ansari, 266.
Anscombe, 210.
Armitage, 84.
Arvesen, 107, 108, 109.
Atkinson, 198.

Baldauf, 112.
Bancroft, 144.
Banerjee, 230.
Barlow, 77, 78.
Barnett, 10, 16, 202.
Bartholomew, 77, 78, 88.
Bartlett, 92, 230, 262, 276, 277.
Barton, 266.
Bassett, 204.
Beale, 250, 254, 255.
Beegle, 251, 253.
Bement, 211.
Bennett, 48.
Berkson, 212, 222.
Bernard, 139.
Berte, 65.
Bickel, 28, 201, 204, 210.
Birch, 226.
Bishop, 207.
Bonferroni, 148.
Bowden, 176, 180.
Box, 18, 80, 81, 89, 90, 91, 94, 135, 140, 142, 149, 197, 198, 201, 208, 264, 267, 268, 269, 271, 272, 274, 276, 277.
Bradley, 266.
Brillinger, 233.
Brody, 163.
Brown, G.W., 83, 93, 203.
Brown, L.D., 73.
Brown, M.B., 269, 273.
Brown, M.L., 232.
Brown, R.A., 81, 91.
Brunden, 139.
Brunk, 77, 78.
Bulmer, 100, 156.
Burbank, 163.
Butcher, 255.

Carroll, 198, 201, 202, 203, 204, 210, 213.
Chacko, 77, 88.
Chan, 228.
Chen, 16.
Chernoff, 11.
Cochran, 84, 92, 110, 128, 163, 242, 253, 263.
Conover, 26, 47.
Cook, 200, 210.
Cornfield, 144, 150.
Cox, 18, 128, 163, 197, 198, 201.
Crawford, 211.
Creasy, 228.
Creger, 65, 113, 161.
Cressie, 274.
Cureton, 26.

David, 263, 266.

Davies, 201.
Deal, 105.
DeGroot, 2.
Deming, 227.
Devlin, 205.
Dixon, D.O., 104.
Dixon, W.J., 31, 35, 55.
Doksum, 201, 202.
Draper, 123, 132, 167, 193, 198, 201, 220.
Duan, 202.
Duncan, 104, 199, 205.
Dunn, 75, 85.
Dunnett, 73, 93, 130, 147.
Durbin, 219, 250.
Dwass, 85, 87.

Efron, 9, 28, 54, 101, 103, 104, 205.
Eilam, 113.
Eisenhart, 118, 263.
Ekbohm, 65.
Eliastam, 65.
El-Sayyad, 228.

Faith, 104.
Feldt, 143.
Fieller, 179, 190, 193, 243.
Fienberg, 207.
Finch, 264.
Finney, 48, 179, 189, 193.
Fisher, 27, 47, 175.
Forsythe, 269, 273.
Francia, 14, 15, 82.
Fraser, 34.
Freund, 266.
Friedman, 65, 137, 141.
Fuller, 211, 231.

Gabriel, 74.
Gastwirth, 35, 36, 37.
Gautschi, 156.
Gayen, 8, 42, 80, 175, 264.
Geary, 7, 8, 42, 80, 243, 264.
Geisser, 142.
Gibbons, 2, 206, 266.
Gleser, 228.
Gnanadesikan, 14, 205.
Gray, 202.
Graybill, 96, 105, 128, 144, 155, 160, 176, 180, 228.
Greenhouse, 47, 142.
Grizzle, 47.
Gross, 31, 32, 204.

Hahn, 174, 177.
Hájek, 267.
Hall, 163.
Hamilton, 181.
Hampel, 28, 30, 32.
Hannan, 35.
Harter, 15, 72, 186.
Hartley, 15, 23, 50, 72, 92, 186, 251, 262, 263.
Hayter, 73.
Hendrickson, 174, 177.
Herbach, 152, 156.
Hettmansperger, 140.
Hill, 208.
Hinkelmann, 105.
Hinkley, 18, 205, 243.
Hoadley, 181.
Hoaglin, 201.
Hochberg, 74.
Hodges, 24, 26, 52, 97, 139.
Hoeffding, 52.
Holland, 207.
Hollander, 64, 84–85, 88, 138, 139, 203, 206, 266, 267.
Hotelling, 146, 175, 176, 180.
Hsu, 48, 62.

Huber, 28, 32, 201, 204, 274.
Hunter, 181.
Hutchinson, 251.
Hutton, 201.
Huynh, 142.

Iman, 85.
Irwin, 47.
Isenberg, 66.

Jacquez, 211.
James, 102, 103, 157.
Jensen, 142.
Johnson, N.L., 113.
Johnson, P.K., 113.
Jonckheere, 87, 88.
Joshi, 104.
Jurečková, 204.
Jutzy, 65.

Kaplan, 110.
Kelly, 229, 232.
Kempthorne, 128, 135.
Kendall, 110, 205, 223, 226, 228, 235.
Kettenring, 205.
Keyes, 163.
Klotz, 50, 98, 152, 267.
Koenker, 204.
Kolmogorov, 14, 82.
Kopecky, 202.
Koziol, 87.
Kraft, 84.
Kramer, 73, 74, 85, 130, 186, 188, 192.
Kruskal, 84, 85,87, 88, 93, 208.
Krutchkoff, 181.
Kurtz, 73.

Lamboy, 181.
LaMotte, 99, 152.
Latscha, 48.
Layard, 107, 108, 109, 199, 205, 272, 273, 274.
Lehmann, 24, 26, 52, 85, 97, 138, 139.
Lev, 52, 207.
Levene, 269, 274.
Lewis, 10, 16, 202.
Lieberman, 11, 181.
Lin, 64.
Lindley, 103, 157, 227, 228.
Lucas, 163.
Lukis, 161.

Mack, 139.
Madansky, 222, 227.
Mak, 228.
Malley, 246.
Mann, 49, 64, 85, 205.
Mantel, 47.
Marsaglia, 243.
Mather, 211.
McLaughlin, 31.
Medeiros, 112.
Mehra, 139.
Mickey, 251.
Mihm, 277.
Miller, J.J., 147.
Miller, R.G., 16, 65, 66, 72, 75, 84, 85, 86, 108, 109, 129, 138, 140, 173, 174, 176, 179, 180, 181, 186, 200, 239, 257, 268, 269, 271, 273, 274.
Milliken, 128.
Milton, 98, 152.
Mohberg, 139.
Mood, 83, 93, 143, 203.
Morris, 101, 103, 104.
Moses, 52, 75, 266, 267, 268.
Mosteller, 253, 273.
Murdock, 260.

Nair, 230.
Neel, 66.

Nelson, 77.
Nemenyi, 84, 85, 87.
Nieto de Pascual, 251.
Nikoskelainen, 66.
Noether, 139.
Northway, 115.
Norwood, 105.

Olkin, 143.
Owen, 19, 20, 23, 50, 72, 138, 186, 206, 263.

Page, 138.
Paulson, 243.
Pearson, 7-8, 15, 23, 35, 42, 43, 50, 72, 80, 186, 199, 206, 262, 263, 264.
Petriceks, 115.
Pierce, 202.
Pitman, 53, 135, 140.
Please, 8, 43, 264.
Pledger, 64.
Pool, 38.
Portnoy, 98, 109, 152.
Pratt, 2, 25, 26.
Pukelsheim, 99, 152.
Putter, 22.

Quade, 85.
Quenouille, 250.

Radtka, 112.
Raffin, 277.
Rahe, 26.
Rand, 161.
Rankin, 125.
Rao, C.R., 98, 152, 175, 252.
Rao, J.N.K., 211, 251, 252, 253.
Rao, P.S.R.S., 110, 251, 253.
Reaven, 257.
Reid, 87.
Ringland, 89.
Rogers, 28.
Rosenblatt, 181.
Rosenthal, 277.
Ross, 251.
Rowell, 143.
Rubin, 35, 36, 37.
Ruppert, 202, 203, 204, 210, 213.

Sarangi, 139.
Satterthwaite, 61, 94, 100, 148, 156, 157, 158.
Scheffé, 57, 60, 71, 74, 75, 80, 85, 90, 91, 94, 100, 127, 130, 135, 144, 146, 148, 150, 156, 173, 181, 246.
Schmitz, 107, 108.
Schou, 142.
Schrader, 140.
Scollay, 255.
Scott, 252.
Searle, 96, 99, 144, 147, 151, 152, 155.
Sen, 203.
Serfling, 63.
Shapiro, 14, 15, 16, 82.
Shorack, 268, 269, 271.
Shrivastava, 230.
Šidák, 174, 267.
Siegel, A.F., 203.
Siegel, S., 266.
Skillings, 139.
Smirnov, 82.
Smith, 123, 132, 167, 193, 198, 220.
Solomon, 263.
Spearman, 206.
Spiegelman, 181.
Shahinian, 115.
Spjøtvoll, 74, 100.
Sprent, 228.
Steel, 85, 86, 87.
Stein, 98, 102, 103, 104, 157.

Stetz, 277.
Stevens, D.A., 66.
Stevens, J.H., 277.
Stigler, 29.
Stoline, 74.
Stuart, 110, 223, 226, 228.
Student, 2.
Swenson, 112.
Szatrowski, 147.

Tamhane, 93.
Terpstra, 87.
Theil, 203.
Tin, 250, 251, 255.
Tukey, 18, 23, 24, 28, 31, 38, 71, 72, 73, 74, 75, 78–80, 81, 85, 88, 91, 108, 112, 126, 127, 129, 130, 131, 134, 136, 144, 147, 148, 150, 160, 186, 188, 192, 253, 266, 273.

Vaeth, 143.
van Eeden, 84.
van Elteren, 139.

Wald, 230.
Walker, 52, 207.
Wallis, 84, 85, 87, 88, 93.
Walters, 143.
Wang, 205.
Watson, 208, 219.
Webster, 251.
Weisberg, 210.
Weissman, 255.
Welch, 55, 60, 61, 64, 65, 93, 135, 140.
Welsch, 201.
Whitney, 49, 64, 85, 205.
Wilcoxon, 22, 49, 63, 65, 84, 85, 87, 205, 268.
Wilk, 14, 16.
Williams, 211.
Williford, 260.
Winsor, 29.
Wolfe, 85, 88, 138, 139, 203, 206, 266, 267.
Wolter, 231.
Wong, 202.
Working, 176, 180.
Wu, 252.

Yates, 46, 47.
Yuen, 31, 55.

Zacks, 98, 152.
Zyskind, 208.

SUBJECT INDEX

A linear $\times B$ interactions, 133.

A linear $\times B$ linear interactions, 134.

A quadratic $\times B$ interactions, 133–134.

Abelson-Tukey approach, 131–135.
  test, 88, 91.

ACD, 38.
  plus adenine, 38.

AHG, 38.

Aligned ranks, 139.

Analysis of covariance, 184.

Analysis of variance, 69–71, 119–120.

APF-test, 271.

Approximate analysis of variance, 124.

Approximate standard deviation for $\hat{x}$, 181.

Asymptotic correlation, 38.

Asymptotic normal distribution of U, 50.

Average rank vector, 84.

Bartlett's test, 262.

Baseline adjustment, 218–219.

Bayes analysis, 2.
  estimators, 98, 152.

Beale's estimator, 250.

Berkson model of a controlled experiment, 222.

Binomial distribution, 19–20.

Bisquare estimator, 31–32.

Bivariate normal model, 242–243.

Biweight estimator, 31–32.

Blocking effect, 33, 110–111, 214–218, 275–276.

Bonferroni inequality, 75.
  intervals, 74–75, 130, 148.
  intervals for contrasts, 76.

Bootstrap method, 28, 54.
  for the transformed correlation coefficient, 204.

Box-Andersen test, 270–272.

Box-and-whisker plots, 4.

Box-Cox method, 197–198.

Box's test, 267–268.

Bradykininogen levels, 113–114.

Brown-Mood median test, 93.

Calibration problem, 169, 177–181, 183–184.

χ^2 statistic, 46.
  for the equality of I proportions, 83–84.

χ^2 test, 14.

Circular serial correlation coefficient, 35.

Cochran's test, 263.

Coefficient of variation, 59, 248.

Concave quadratic regression function, 195.

Conditional variance of U, 51.

Confidence interval(s), 4, 41.
  for the true difference in location of two populations, 52–53.
  for variance components, 156.

Contaminated normal distributions, 10.

Continuity correction, 46–47.

Contrast, 75–76.
  agents, 162–163.

Control and experimental, 40.

Correlation coefficient, 169, 199.

Counting method, 49.

Cross-classification model, 151.

Delta method, 58–59.

Dependence, 32–37, 63–64, 94–95, 110–111, 141–143, 150, 159, 214–220, 234, 254, 275–276.

Dependent variable, 169.

Difference model, 164.

Digital subtraction angiography, 162.

Distribution theory for the sum of squares, 70–71.

Empirical Bayes estimators, 102–104.
Epstein-Barr virus, 65–66.
Errors-in-variables model, 167, 220–234.
Estimation of individual effects, 157–158.
of variance components, 96–99, 151–155.
Expected mean squares, 121.
Experimental and control, 40.
Exponential distribution, 9.
Extra period Latin square design, 162–163.

F test, 71.
F statistics, 121–122.
Fibrinogen levels, 161–162.
Fieller interval(s), 190, 193.
Fisher's exact test, 47.
Fixed effects, 67–68, 69–95, 118–143.
model, 118.
Friedman's rank test, 137–138.
Functional relationship between x and y, 222.

Glucose tolerance test, 257.
Gold standard, 221.
Graybill and Bowden band, 176.
Graybill model, 144–148.
Group mean method, 230.

Hartley's test, 262–263.
High endothelial venules (HEV), 255.
Hodges-Lehmann estimator, 24, 39, 52, 97.
Hodgkin's disease, 113–114.

$I > 2$ sample variance problem, 262–263.
Independent variable, 169.
Individual population means estimates, 101–104.

Intercept, 168.
Interquartile range (IQR), 275.
Intraclass correlation coefficient, 110.

Jackknife, 108–109, 148, 149, 159, 232–233, 250–252, 272–274.
Jackknifing the transformed correlation coefficient, 204.
James-Stein estimator, Lindley form of, 103.

Kanamycin, 238–240.
Kendall's coefficient τ, 205–206.
Klotz-Milton-Zacks estimators, 98.
Kolmogorov-Smirnov test, 14.
Kruskal-Wallis rank test, 84, 93.
Kurtosis, 6, 7, 42, 43, 82.
test, 15.

L-estimator, 28–31.
Large sample theory, 5.
Large sample variances of $\hat{\alpha}$ and $\hat{\beta}$, 228–229.
Least squares estimates, 169–171.
Levene s test, 269.
Likelihood, 2–3.
Likelihood ratio test, 1–2, 37, 64, 69–71, 77–78, 99, 119.
Linear contrast, 79.
Linear-2 contrast, 79.
Linear-2-4 contrast, 79–80.
Logarithmic transformations, 16–17, 44–45, 82.
Log normal distribution, 249.
Log-probit paper, 13–14.

m-dependence, 34.
M-estimator, 28–29, 31–32.
MAD, 32.
Mann-Whitney form of the Wilcoxon statistic, 49.
Mann-Whitney *U* statistic, 64.

Maximum likelihood estimates, 77, 97, 169–171, 182, 187, 224–226, 243, 246.

Mean squares, 69.

Mean sum of squares, 69.

Median, 12, 18, 20, 39, 83.
absolute deviation (MAD), 275.
based estimators, 203.
χ^2 test, 65.

Method of moments estimators, 148, 151, 153–154, 227, 231.

Minimum norm quadratic unbiased estimators (MINQUE), 98.

Mixed effects model, 118, 143–150.

Models I, II, and III, 118.

Monotone alternatives, 76–80, 131–135, 148.

Multiple comparisons, 71–76, 129–131.

Multiplicative model, 164–166.

Multisamples: general intercepts, 184–191.

Multisamples: zero intercepts, 191–193.

Nested designs, 157.

Nested model, 151, 153–155.

Nonlinearity, 193–198, 231.

Nonlinear regression analysis, 193–194.

Nonnormality, 5–32, 80–89, 105–109, 135–140, 149, 158–159, 199–207, 231, 264–275.

Nonparametric techniques, 19–28, 45–54, 82–99, 137–140.

Normal linear model, 168–193.

Normal scores test, 26.

Normal theory, 1–5, 143–148, 150–158, 259–263.
one-way classification, 69–80.
two samples, 40–41.

Notation: $\approx$, 20.
cdf, 5.
χ^2_ν, 61.
$\chi^2_\nu(\delta^2)$, 71.
df, 2, 70.
$\sim$, 56.
F_{ν_1,ν_2}, 99.
$F_{\nu_1,\nu_2}(\delta^2)$, 122.
$F_{\nu_1,\nu_2}(\delta_1^2,\delta_2^2)$, 126.
MS, 69.
$N(\mu,\sigma^2)$, 1.
$\Phi(\cdot)$, 13.
$q_{k,\nu}$, 72.
$SD(y)$, 59.
SS, 70.
s^2, 2.
s_x^2, 244.
s_y^2, 244.
s_{xy}, 244.
$|m|_{k,\nu}$, 174.
$t_\nu(\delta)$, 78.
VDT, 70.
$\bar{y}$, 2.
$\bar{y}_{i\cdot}$, 126.
$\bar{y}_{\cdot j}$, 126.
$\bar{y}_{\cdot\cdot}$, 126.
$\bar{y}_{ij\cdot}$, 120.
$\bar{y}_{i\cdot\cdot}$, 120.
$\bar{y}_{\cdot j\cdot}$, 120.
$\bar{y}_{\cdot\cdot\cdot}$, 120.
$\bar{y}^*_{i\cdot\cdot}$, 124.
$\bar{y}^*_{\cdot j\cdot}$, 124.
$\bar{y}^*_{\cdot\cdot\cdot}$, 124.
$\xrightarrow{d}$, 5.
$\xrightarrow{p}$, 5.

One degree of freedom for nonadditivity, 127.

One sample: general intercept, 168–181.

One sample variance problem, 259–260.

One sample: zero intercept, 181–184.

Opacification index, 162–163.

Optical density, 236–238.

Ordered values, 11.

Outliers, 9–10, 12–14, 18–19, 44, 45, 199, 231, 253, 274.

Overall mean, estimate of, 104, 158.

Oxygen toxicity, 115–116.

P value, 2, 3, 8, 19–20.

Percent error, 59.

Percentage change, 241.

Permutation test, 27–28, 88–89.

Pitman's permutation test, 53–54.

Pooled sample variance, 41, 262.

Power family, 17.

Prediction problem, 169, 175–177.

Predictor variable, 169.

Probit paper, 10–11.

Probit plots, 10–12, 39, 44, 45, 82, 107, 136.

Product-moment correlation coefficient, 35, 171.

Pseudo-values, 108.

Q-Q plotting, 14.

Quadratic effects, 132.

Random block effect, 63.

Random effects, 68, 95–111, 150–159.
  model, 118.

Ranking, 49.

Ratios, 241–254.

R-estimators, 29.

Regression model, 166–167, 168–220.

Relative change, 241.

Relative potency, 189–191, 192–193.

Repeated measure designs, 142–143.

Resistant, 10.

Response variable, 169.

Robust estimation, 28–32, 54–55, 89, 140.

Robust for efficiency, 9.

Robust for validity, 9.

Sample serial correlation coefficient, 35.

Satterthwaite's χ^2 approximation, 61–62, 100–101.

Scheffé intervals, 74, 130.

Scheffé model, 144.

Sequence effect, 33–37.

Serial correlation, 34–37, 94–95, 111, 142, 219–220.

Serial correlation coefficient, 63.

Serial dependence, 276.

Shapiro and Wilk test, 16.

Shapiro-Francia statistic, 14–15.

σ, estimate of, 12.

Single missing value, 128.

Signed-rank statistic, 23, 36–37.
  alternative representation, 23.

Signed-rank test, 39.

Sign statistic, 36–37.

Sign test, 18–22, 39.

$\sin^{-1}\sqrt{\hat{p}}$, 17.

Skewness, 6, 7, 42, 43, 82.
  test, 15.

Slope, 168.

Slope ratio assay, 193.

Splenectomy, 161–162.

Square root transformation(s), 17, 44, 82.

Standard error, 4.

Standard line, 180.

Steel-Dwass test, 86.

Stein estimators, 98.

Structural relationship between x and y, 222.

Studentized maximum modulus distribution, 174.

Studentized range distribution, 85.

Studentized range test, 91.

Sums of squares, distribution theory for, 70–71.
Survey sampling, 242.

t distribution, 6.
t statistic, 5, 6-9, 41.
t test, 37, 39, 65.
  with winsorized standard deviation, 39.
$\tanh^{-1} r$, 17.
Tests for variance components, 155–157.
Test of equal slopes, 185–186.
Thermodilution cardiac output (TDCO), 277.
Ties, 21–22, 24–26, 28, 50.
Ties correction, 85.
Tin's estimator, 250.
Transformations, 16–19, 44–45, 58–60, 82, 92–93, 108, 137, 141, 150, 159, 196–198.
Transformed correlation coefficient, 175.
Trimmed mean, 29–31, 39, 54–55.
Trimmed regressions, 203–204.
Trimmed t statistic, 55.
Tritiated thymidine, 39, 115–116.
Trunk flexor muscle strength, 112.
Tukey-Kramer intervals, 73, 188, 192.
Tukey studentized range, 147.
  intervals, 129.
  test, 72.
2×2 contingency table, 46.
Two sample median test, 45–48.
Two sample t test, 64.
Two sample Wilcoxon rank test, 63.
Two sample variance problem, 260–261.
Two-tailed exponential distribution, 6.

Unequal variances, 56–63, 89–94, 109–110, 140–141, 149–150, 159, 207–214, 233–234, 253–254.
Uniform distribution, 6.
Uniformly most powerful invariant test, 98.
Uniformly most powerful similar test, 98, 155.
Uniform minimum variance quadratic unbiased estimators, 96, 151.
Uniform minimum variance unbiased estimators, 96, 151.
Unpaired data, 246–249.

Variances, 259–276.

Weighted least squares analysis, 210–214.
Welch's t' statistic, 55.
Welch's t' test, 60, 64, 65.
Wilcoxon rank test, 49, 65.
Wilcoxon signed-rank test, 22–26.
Winsorized mean, 30.
Winsorized variance, 29–31.
Working-Hotelling band, 176.